POLICY AND PRACTICE IN EDUCATION

NUMBER ONE

EDUCATION AND THE SCOTTISH PARLIAMENT

POLICY AND PRACTICE IN EDUCATION

1: Lindsay Paterson, *Education and the Scottish Parliament*
2: Gordon Kirk, *Enhancing Quality in Teacher Education*
3: Nigel Grant, *Multicultural Education in Scotland*

POLICY AND PRACTICE IN EDUCATION

EDITORS

GORDON KIRK *AND* ROBERT GLAISTER

EDUCATION AND THE SCOTTISH PARLIAMENT

Lindsay Paterson

DUNEDIN ACADEMIC PRESS

EDINBURGH

Published by
Dunedin Academic Press Ltd
8 Albany Street
Edinburgh EH1 3QB

ISBN 1 903765 00 5

British Library Cataloguing in Publication Data

A catalogue record for this book is available from the British Library

Typeset by Trinity Typing, Coldstream
Printed in Great Britain by Polestar AUP Aberdeen Limited

CONTENTS

EDITORIAL INTRODUCTION

Education is now widely regarded as having a key contribution to make to national prosperity and to the well-being of the community. Arguably, of all forms of investment in the public good, it deserves the highest priority. Given the importance of education, it is natural that it should be the focus of widespread public interest and that the effectiveness and responsiveness of the educational service should be of vital concern to politicians, teachers and other professionals, parents and members of the general public. If anything, the establishment of Scotland's parliament, which has already affirmed education as a key priority, will witness an intensification of public interest in the nature and direction of educational policy and the changing practices in the schools. This series of books on *Policy and Practice in Education* seeks to support the public and professional discussion of education in Scotland.

In recent years there have been significant changes in every aspect of education in Scotland. The series seeks to counter the tendency for such changes to be under-documented and to take place without sufficient critical scrutiny. While it will focus on changes of policy and practice in Scotland, it will seek to relate developments to the wider international debate on education.

Each volume in the series will focus on a particular aspect of education, reflecting upon the past, analysing the present, and contemplating the future. The contributing authors are all well established and bring to their writing an intimate knowledge of their field, as well as the capacity to offer a readable and authoritative analysis of policies and practices.

The author of this volume, the first in the series, is Lindsay Paterson, Professor of Educational Policy at The University of Edinburgh.

Professor Gordon Kirk
Faculty of Education
The University of Edinburgh

Dr Robert T D Glaister
School of Education
The Open University

ACKNOWLEDGEMENTS

I am very grateful to David McCrone for making numerous insightful helpful comments on a draft of the book, and to Bob Glaister and Go Kirk for the invitation to write it.

References in the main text have been kept to a minimum, and are ma confined to providing precise sources for quotations and for statist information. The guide to further reading at the end indicates the gen sources on which the book is based.

CHAPTER 1

INTRODUCTION

Education cannot help being at the centre of the concerns of the Scottish Parliament. For over a century now, there has been a strengthening belief that a parliament could make better policy for education than the unreformed Union has managed, and could link that efficiently to social needs. This short book is an essay on the debates which are likely to ensue. It is about the politics, government and culture of Scottish education, rather than directly about what goes on inside the classroom. It concerns what a Scottish Parliament is intrinsically about — democratic debate — rather than educational changes that could, in principle, take place under any governing system. Of course, claims about the curriculum, about examinations and about standards appear frequently in the political discussion of education, but it is as political topics that these are dealt with here — what impact the Parliament might have on how they are debated and on how resolutions of the controversies are found. The same is true of the thorny old question of whether Scottish education is the best in the world or appallingly mediocre: this book makes no attempt to assess such extravagant claims, but it notes them because they are such potent elements of the political debates.

The debates are taking place because the optimism is so entrenched, and has been accumulating for such a long time. Typical is a 1989 report from a committee of eminent writers on education and culture, published shortly after the Constitutional Convention had started the process that has now led to the creation of the Scottish Parliament. The committee concluded that

> if Scottish education is to meet the challenge of the future, to develop in response to the needs of the Scottish people, it has to be under the control of the Scottish people. (SCESR, 1989, p. 32)

That is broadly representative of the vaguely nationalist sentiment which has pervaded educational debate throughout the period when Scottish self-government has been dominating political discussion more generally. Further back, in 1968, the noted philosopher H. J. Paton claimed that 'in the field of education, as elsewhere, Scotland needs genuine autonomy if she is not to be unworthy of her past' (Paton, 1968, p. 250). Further back still, in 1925, the Independent Labour Party MP James Maxton argued at the annual congress of the teachers' trade union, the Educational Institute of Scotland, that a separate Scottish Parliament was needed so that Scottish national ideals could shape the education system (Maxton, 1926). Commenting in the union's journal shortly afterwards, the nationalist poet Hugh MacDiarmid agreed, although he argued also that these national ideals had been corrupted by 'a preponderantly English legislature', and required fundamental renewal before a Scottish Parliament could make much of a difference (Grieve, 1926, p. 252). Even in the nineteenth century, at the height of Scotland's absorption into Britishness and Empire, a nationalist interpretation of educational problems was beginning to take shape. Lord Reay, newly elected as rector of St Andrews University in 1885, believed that

> if we had a Scottish Parliament sitting in Edinburgh, I have no doubt that the organisation of the Universities would be the first number on the legislative programme. (quoted by Anderson, 1983, p. 269)

That history of enthusiasm for a Scottish Parliament among educationalists has also now permeated Scottish popular attitudes. Two out of three people preparing to vote in the 1997 referendum believed that a Parliament would make Scottish education better, and a mere one in twenty-five supposed it would make it worse. That (and similar beliefs for health and social welfare) strongly influenced how people actually voted (Brown et al, 1999, pp. 118–22). Now that the parliament is in place, it is then not surprising that a piece of proposed educational legislation has headed the programme of the coalition government.

These views, however, are not the only reason to expect that debating a Scottish Parliament would inevitably entail debating the future of education, and thus to expect that the coming of the Parliament would have profound consequences for education. Scottish identity itself (whether nationalist or unionist) has long been held to be intimately linked to Scottish education — a belief that goes right back through the Union of 1707 to the Reformation's attempt to set up a school in every parish and to James IV's legislation of 1494 to enforce school attendance on the sons of significant landowners. There would have been wide agreement in the nineteenth century, for example, with the belief of James Lorimer, professor of public law at Edinburgh University, that Scotland's national distinctiveness was 'not political, or even institutional, but social, and, above all, intellectual' (quoted by Anderson, 1983, p. 61).

Claims that Scottish identity depended on its education system — at least in part — have been intimately mixed up in the debate about self government in the last three decades. John P. Mackintosh — professor of politics, Labour MP and ardent home ruler — argued in 1974 that

> the uniting influence of common systems of education, local government, law and religion [has] imbued all those who have been to school, lived and worked in the country with a sense of being Scottish. (Mackintosh, 1974, p. 408)

That view was not confined to advocates of self government. During the heated debates in the 1980s over the various pieces of education legislation from the Conservative government, all sides argued that Scotland had unique educational traditions. Conservatives such as Malcolm Rifkind and Michael Forsyth both attempted to reform the system according to the ideas of the New Right but also in ways that — they believed — would be quite consistent with Scottish traditions (Paterson, 1994, pp. 1–2).

Thus there has broadly been a belief that the essence of the nation depended on its education system. If a Parliament is partly an expression of national identity, and if education is at the heart of that identity, then the coming of a Parliament must surely — it is widely believed — affect the cultural role which education is imagined to have.

That, moreover, is a consequence of a much more general feature of nationalism in the last century and a half. Every national movement has had an educational programme. Nationalists have believed that the bonds of community can be re-created by forging a common identity through education. They have believed that a renewed national identity can be brought into being by education, and also that only such a common identity could give political legitimacy to the new state which the nationalist project sought to build. Similar views were held by the various imperial powers which resisted the nationalists, and so there have been numerous instances of conflict over the cultural and political purposes of education. A classic instance is in Ireland, where the British government reformed the education system in 1831 in order to try to create a culture that would be loyal to the empire. In due course, that provoked a response from nationalists such as Patrick Pearse, a leader of the Easter Rising and by profession a teacher in his own independent school. The schools, he believed, should be used to create a new national character, free of what he called 'the English education system' and its tendency to 'mak[e] slaves of us' (Lyons, 1973, p. 89). The new Free State after 1922 then attempted to use the schools to create a new national identity — what one professor called 'the form and kind of civilization developed by a particular people' (Brown, 1985, p. 48).

Through numerous reiterations of such views in all varieties of nationalist programme there has emerged the common assumption that national self-government and national educational development are intimately related to each other, even where social, historical and political circumstances are vastly different from those which prevailed in places such as Ireland in the nineteenth century. From nationalists of the right and of the left, from moderate home rulers and from radical secessionists, there has been agreement on the social and cultural importance of education. No wonder, then, that somewhere in the back of most people's minds in Scotland at the beginning of the new century is a sense that the education system must respond to the new political era.

This book is about Scotland, but the themes it deals with are by no means unique to Scotland: in some form, they are common across the developed world, and in many respects also further afield. Scotland may have a unique combination of common themes, but so does every national system of education: at least since the beginning of the industrial revolution, national distinctiveness in education has lain much more in a nation's peculiar way of absorbing international currents of thought and development than in any native genius.

For our account of the politics of education in the newly self-governing Scotland, five common themes can be found linking this experience to that elsewhere, and so helping to explain the Scottish situation by reference to international processes. We will come across these themes recurrently in later chapters. The first is the fundamental role of education in shaping national identity in every industrial society: that is ultimately why nationalists have paid education so much respect. The best-known explanation of this is from Ernest Gellner's *Nations and Nationalism* (1983). He argued that mass systems of education were required because industrial society depended on massive amounts of social mobility. There was no longer the rigidly separate castes of previous social systems, and even if, in fact, people could not move freely up the hierarchy of social class, there had to be the illusion of social mobility to maintain the moral acceptability of the social order. Education systems based on individual ability and effort — meritocracy, as it has come to be called — provide that moral sanction, because they seem to place social inequality on the morally acceptable basis of differences in educational attainment. To achieve this social fluidity, most training had to be generic, not specialised, so that people could be free to move: everyone had to be trained in 'literacy, numeracy, basic work habits and social skills' (p. 28). This need for common education was reinforced as most work came to involve the manipulation of meanings, rather than things: this is reminiscent of Benedict Anderson's dictum that nations are 'imagined communities', created by books, newspapers, and other ways of sharing meanings. A common education of that sort required a system of common governance — a common politics of education: taking issue with Max Weber's definition of the state, Gellner proposes that 'the monopoly of legitimate education is now more important than is the monopoly of legitimate violence' (p. 34). And so education now confers identity. Loyalty is to a culture, and culture is largely transmitted by schools, not mainly custom, or folk tradition, or stable community.

Despite this last point, Gellner strangely does not make much of the more explicitly political roles which education came to play in the modern era. Others have filled that gap. R. D. Anderson (1995, p. 193), for example, suggests that national education systems consolidated the nation state, smoothed the introduction of mass political democracy — offering a secular religion, according to Durkheim — and allowed for the peaceful integration of the industrial masses (who were in danger of slipping off into socialist radicalism). So education could hardly be more important in the modern era: it is the key to economic efficiency, cultural homogeneity and political legitimacy.

The second common experience in which Scotland has shared expands on part of Gellner's argument. In almost all nations, education has been built up by the state, although with varying degrees of central prescription. The major exception is England, argued by Andy Green (1990, 1997) to have been exceptional in educational development — as in much else — because it was the first nation to make the transition to industrial modernity. This is a familiar argument, but it also serves to situate Scottish education closer to mainland European experience than to that of its immediate neighbour. R. D. Anderson puts it well: the view of public policy for education in nineteenth-century Scotland was close to the 'German idea of the ethical state' (1992, p. 69).

Scotland then found it relatively easy to make the transition to a system of mass education for mass democracy, the third common theme. The extension of the franchise in almost all European countries in the early decades of the twentieth century created a common pressure for mass education as a social right, and nervous political elites also came to believe that widespread education was needed to enable the new electorates to govern the nation wisely: 'we must educate our masters' is the misquotation that sums this up, neatly encapsulating the sense of a rather more convoluted statement from the English Liberal politician Robert Lowe in 1867, preparing for the introduction of a state system of elementary education there in 1870 (Dyson and Lovelock, 1975, p. 201). These new mass systems were at first rigorously meritocratic, at least according to intent, selecting children for successive stages by stringent tests of ability. But each of these stages eventually came to be seen as a right, not as a privilege to be earned: first elementary education, then secondary, then secondary of a common academic type (comprehensive education), then secondary beyond the middle years of adolescence, and finally — more recently — education beyond school and even education throughout the lifespan. In Scotland as elsewhere, then, access to education has come to be attached to political campaigns for social justice, and the demands of justice seem to be insatiable, insofar as they seem to be attached to successively more advanced levels of study. Putting in place more and more education has not been simply a matter of responding to economic imperatives, although these have been important: thus promoting lifelong learning has the economic advantage of developing a flexible workforce. The extra educational opportunities which these economic changes sanction are expected to be distributed more equally than the opportunities in a system that trains only a few people to very high levels.

The political system which created this mass system was also, however, firmly not based on popular participation, beyond voting. It was governed by what has been called technocracy, rule by experts — in education, not so much teachers in general, but rather the educational elites in the schools inspectorate, the central and local educational bureaucracies, the culturally influential professoriate in the universities, and the network of specialist advisory and executive committees which politicians set up as a means of distancing themselves from difficult or controversial decisions. The sheer power of these elites may have been particularly evident in Scotland, where there was no indigenous parliament, but they were at the heart of educational policy making everywhere.

But then — the fourth common theme — there was a reaction against that, in the individualistic assertion of the 1960s and after. This has now profoundly influenced the character of educational governance across Europe and North America, through such widely adopted policies as the devolution of power to schools and other educational institutions, the placing of more power in the hands of parents and (for older groups) students, and the tighter monitoring of professional activities in the name of accountability to consumers rather than to the community. In a sense, though, this individualism is also a consequence of a much more educated society. Better-educated people demand more rights, are less respectful of authority, are more tolerant of diverse lifestyles and beliefs, and want more influence over how public services such as education are run.

So this common theme is linked not only to the new right governments of politicians such as Margaret Thatcher, but also to the gaining of the new Scottish Parliament itself. That is the Scottish version of the widespread scepticism of state structures. The campaigning for a Parliament has been accompanied by enthusiasm for citizenship education, for participatory decision making and for renewing the moral basis of democracy; all these themes can be found in many other places, including England, and all depend on a renewal of the political role of education. Moreover, despite the rhetoric of the years of new right dominance of political debate in the UK and many other places, education continues to be regarded as a public good everywhere: again Scotland is not unique. No new right government actually privatised more than fairly peripheral parts of education. They also all presided over a growing educational respect for cultural pluralism, despite the ostensible aim to use curricula to reinforce fairly homogeneous ideas of national culture. In many respects, indeed, the period from the 1970s to the 1990s saw a redefinition of national identity in western Europe and north America as intrinsically multi-cultural, and education has been as central to that as it was to the attempts to build homogeneous national cultures a century earlier.

One of the reasons all this could happen despite new right rhetoric brings us to the fifth and last common theme: the autonomy of civil society from the state. Despite the political aspirations, quangos and inspectorates run the systems according to their own criteria, and teachers continue to have considerable independence inside their own classrooms. Because civil society has thus seemed to many writers — both political and academic — as morally more acceptable than corrupt and distant politicians, there has emerged the view that civil society can be the site for re-moralising politics. That theme has been very common in Scottish debates about national self-government, but it can be found also in many other societies. On the other hand, the proposition that civil society can renew the state seems to contradict the autonomy in which civil society's claim to moral purity lies. Again, that tension is common, and the search for non-coercive and unalienated forms of politics through civil society has been a uniting feature of democratic renewal in central and eastern Europe, movements for regional autonomy in western Europe and north America, and projects of constitutional reform in the old European states. Whether it is possible to bridge that gap between civil society and the state has become one of the most important political questions of our times, or rather we have renewed a question that stretches back at least to the eighteenth century. This fifth common theme is at the heart of the book's argument: the future politics of Scottish education in the context of the new Scottish parliament is a specific national instance of a general dilemma facing all developed democracies.

Before we get to that, though, we examine, in Chapter 2, the historical origins of the sense that Scottish education is at the heart of all sorts of Scottish debates. What has being at the core of Scottish identity meant? Debating Scottish education cannot help debating Scottish government because serving the nation is bound to raise issues of the legitimacy of Scottish government. The claim since the Reformation and earlier that Scottish education is national property was renewed as Scotland came to share in the mass systems of education that grew to accompany mass democracy. The central issue which has then been raised for Scottish government is whether the merely managed autonomy that Scotland has had in the Union could ever be enough to realise the most radical implications of attempting to be public. Could a system of government dominated by civil servants and selected professional advisers rather than by indigenous elections ever be enough to respond to popular pressures for social justice through education?

That is why education has always featured in the arguments for and against self-government, as is examined in greater detail in Chapter 3. On the one hand — as we have already noted — there grew the belief that only a Scottish Parliament could ensure that education served the nation truly. Indeed, as the Scottish self-government movement gradually came to be aligned almost wholly with the Scottish left (and as the left almost entirely came to support self government), the conflicts between managed autonomy and democratic assertion were focused on the issue of social justice. It came, in particular, to be believed that social justice was an age-old aspiration of Scottish education, and that the only way to realise it was through a popularly elected national Parliament. On the other hand, from the opponents of a Parliament came the argument that it could threaten the quality which the leading parts of the teaching profession had given education through their astute management. That argument between self-government movement and venerable institutions seemed to recede as the campaigning for a Parliament reached success, but is bound to return now as conflict between the Parliament and the educational institutions while they compete for influence over the system's direction.

However, despite the argument in Chapter 3, the relationship between the Parliament and the education system will not be resolved in the abstract: it will depend on what the Parliament does in key aspects of educational policy. The Parliament does not start with a blank sheet, and a great deal of its involvement with education — as with other areas of social policy — will be to maintain trends that are already evident. Chapter 4 examines that legacy, arguing that there are four particularly influential features. The first is the simple fact of expansion. The second is the comprehensive principle as the basis of expansion in Scotland — a sense, at least since the 1960s, that expansion is worthwhile because it furthers social justice. The third is an incipient dissatisfaction with the quality of the education system. And the fourth is the democratisation of society and culture that is as thorough in Scotland as almost anywhere else, but which the Scottish political majority have barely faced up to. These democratising changes have been presented in Scotland as coming from the political right, and have been responded to managerially rather than through a well-developed political programme that relates them to Scottish traditions. A self-governing political process will not now be able to avoid the resulting ideological omissions.

In conclusion, it is argued in Chapter 5 that there will be broadly two versions of the new politics of education. On the one hand, the Parliament could work closely with the existing institutions — the pillars of Scottish education that governed it within the Union. One reason why the Parliament might follow this route could simply be the necessity to get things done — to keep the system going. Also working in the same direction would be the inexperience of many of the new members of the Parliament, especially in the face of professional advisers such as the schools inspectorate. Defence of educational professionals' right to run the system is likely also to come from people who were still hostile to a Parliament in 1997: the referendum result was too clear-cut for them to maintain their opposition, but they will question the Parliament's right to interfere with Scotland's traditions. The Parliament could be inclined to interpret the enthusiasm for consensus that has been evident since the referendum as an invitation to be cautious about any radical reforms that had not gained the assent of the existing policy communities. This route might tend to draw the Parliament into civil society. It would frustrate any project for radical reform, but might encourage slow change that was permanent because it had been reached by consensus. Above all, it would represent a victory for the safe traditions of civic Scotland.

On the other hand, a quite different route is available. Along this, the Parliament takes on ancient Scottish civic institutions such as education, and forces them to democratise, to open up, and to renew the ideals of social justice which underlay the founding of the welfare state. Instead of working by consensus with influential professional groups in civic Scotland, this route would involve the Parliament in provoking popular participation against their conservatism. That would include participaion by less influential professionals (such as teachers) against the policy community — the inspectors, quango members and civil servants who have shaped the system in the unreformed Union. The outcome would be much stormier than on the first route, and might be a complete failure of reform to be accepted. But it would be bound to involve an intense public debate about education and its role in Scottish identity.

In short, this book is an essay with an essentially simple thesis. On the one hand, Scottish education has been regarded as largely autonomous of the UK state throughout the period of the Union. Apparent threats to that perceived autonomy have always been resented, and the most recent such threats — from the governments of Margaret Thatcher and John Major — are among the strongest reasons why we now have a Scottish Parliament at all: people voted emphatically yes in the 1997 referendum in part to protect the integrity of their education system. This autonomy of Scottish education is one particular instance of the autonomy of civil society from the state throughout the developed world. On the other hand, the very belief that Scottish education has a right to govern itself is bound to cause the Parliament itself awkward political difficulties. Civic institutions do not suddenly come to love the state just because there has been a change of politicians, and yet the Parliament, being popularly elected, will have a capacity for decisive political action which the Union has lacked for many decades. However important the legislation which the Parliament may pass, it will be discussions about the sources of legitimate authority which will mark its most lasting impact on Scottish culture. The resulting tensions between Parliament and civic Scotland will not only shape the character of the nation for many years to come. Because these conflicts can be found in all democratic societies now, their working out in Scottish education over the next few decades will also provide a test-case of civil society's role in democratic renewal.

CHAPTER 2

EDUCATION AND THE CIVIC AUTONOMY OF SCOTLAND

Accounts of Scotland's place in the Union commonly say that it has depended on three sets of institution: the law, the church, and education. Sometimes local government is added, and sometimes also — for the twentieth century — the media. But education occupies a central role not granted to, say, housing, social services or even medicine. This Chapter summarises what that has entailed. What has being a pillar of identity meant for Scottish education? How did that change as the role of the central state in education grew in the twentieth century? How has the resulting politics of education shaped Scottish politics more generally?

The argument of the present Chapter is that, throughout the three centuries of Union, the system was felt to be self-governing. Both by educationalists and by the people of Scotland more generally, it was believed that education could manage its own affairs, including responding to problems as they arose. Frequently that might entail adapting English or other international ideas, but the important point was that the borrowing was freely chosen, and was chosen in Scottish terms. This is why there is such a strong sense that the education system is the property of the Scottish people, and it is why such political passions are roused when it seems to be threatened from outside or by political philosophies that are believed to come from outside.

These themes from Scottish educational history provide the main context for the role of education in the debates about self-government (which we review in the next Chapter) and for the educational legacy which the Parliament inherits (discussed in Chapter 4). The present Chapter is no substitute for a full history (for references to which, see the guide to further reading at the end): it can do no more than outline why Scottish education has been felt to be autonomous.

Education and Autonomy: 1707–1872

Tracing the autonomy of Scottish education to the Union of 1707 is in some respects an obvious starting point: educational independence was explicitly guaranteed by associated legislation in autumn 1706 which was then incorporated into the Treaty (Daiches, 1977, pp. 145–6). But more important than that in itself was the main purpose of that associated legislation: the guaranteed independence of the presbyterian church. It was the church which ran the parish schools, and so it became in effect the national Parliament so far as school education was concerned. The distinctiveness of Scottish local government also was important for schooling in the large burghs, where the councils were responsible for schools that would have been a parish matter in the landward areas.

The parish system dated in aspiration from the Scottish Reformation of 1560, and in particular from the *First Book of Discipline* compiled by commissioners who were led by John Knox. He followed Calvin's teaching in hoping to place a school in each parish. The system was consolidated by an Act of the Scottish Parliament of 1696, and by the middle of the seventeenth century was largely in place. At that time, Scotland had one of the most developed systems of popular schooling in the world. In the words of one historian of Scottish education between 1760 and 1830,

> towards the close of the period, ... there can be little doubt that the machinery for producing a highly literate population in Scotland was well in place, and indeed had been for some time. (Withrington, 1988a, p. 185)

By the middle of the nineteenth century, that promise had borne fruit, with Scottish literacy levels indeed among the highest in Europe. In 1871, for instance, 89% of Scottish men could write their name, and 79% of Scottish women. This was a similar level to those in Prussia and Scandinavia, some 15 percentage points better than in England, and about twice the level in southern Europe (Cipolla, 1969)

The important point for our purposes is the autonomy: this extensive system of parish schools was the product of an independent church and of legislation inherited from before the Union. By and large, the same can be said of the response which the system made to the new problems posed by industrialism and the rapid growth of towns and cities. Difficulties in the parish system were already being noted in the early years of the nineteenth century. Siting a school in every parish was no longer adequate when some parishes had grown enormously and some new towns — such as Coatbridge — did not even have a parish church.

The first response came from the evangelical wing of the church, led by Thomas Chalmers. He believed that the system could be renewed on its own terms, and indeed that establishing schools in urban areas could help to re-create the communities that industrialism had destroyed:

> The ties of kindliness will be multiplied between the wealthy and the labouring classes, ... the wide and melancholy gulf of suspicion will come at length to be filled up by the attention of a soft and pleasing fellowship. (quoted by Withrington, 1988b, p. 48)

His was the most sustained attempt to set up schools that would still be run by the church and therefore still be independent of the state, and he inspired followers such as Thomas Guthrie whose 'ragged schools' were intended to teach (and morally reform) children who spent the day running around the streets (Anderson, 1995, p. 95). Ultimately these projects failed to provide a system of schooling in the industrial areas because the working class were just too poor to support their own schools, and middle-class philanthropy was not enough to meet the need. Moreover, the capacity of the church itself to take on the system was severely weakened when it split in two at the Disruption of 1843. Chalmers and the evangelicals left to form a Free Church, but — although their achievement in setting up schools in about 400 of the 900 parishes was impressive — neither they nor the continuing established church was able to provide what was required.

The only alternative was aid from the state, as concluded by, for example, David Stow, founder of the teacher training college in Glasgow. He believed it necessary to advocate

> large government grants for the moral and intellectual training of the young, [because] otherwise the people would never educate themselves, and ... the private subscriptions of the wealthy would fail in providing the requisite funds for that purpose. (Stow, 1847, p. 2)

Guthrie and many other evangelicals accepted state aid on these grounds (even though still claiming the legacy of Chalmers, who died in 1847). The supporters of the established church had their own reasons for favouring a national system: in the language of a highly influential pamphlet by George Lewis in 1834, they inclined to argue that Scotland was 'half educated' because half the children were not in Church of Scotland schools. Such a convergence of interest — along with that of secularists, who eschewed church influence altogether — produced wide agreement in favour of a national system, although still a great deal of disagreement over its exact form. That disagreement postponed acceptable legislation until 1872, but did yield the very thorough investigation of the system carried out in the 1860s by an official Commission headed by the Duke of Argyll. He was a strong advocate of the public character of the Scottish tradition — of the sense that the schools belonged to the nation — arguing in the House of Lords in 1869 that

> the parochial system in Scotland was founded by John Knox, who laid down the principle, which has never faded from the popular mind in Scotland, that it is the duty and function of the state to insist upon the education of the people. (quoted by Anderson, 1995, p. 64)

So the national system of elementary schooling established in 1872 can be reasonably argued to have been the outcome of autonomous campaigning responding to failures in the previous system of autonomous schooling that was inherited from the Union. We return to the subsequent evolution of the 1872 system; what matters at the moment is the firm sense that this embodied a national consensus, argued over thoroughly during the previous three decades. It was, no doubt, a consensus modified by what was going on in

England at the same time — the much more bitter arguments around the Education Act of 1870. But it was a Scottish consensus nonetheless, using parallel English developments to obtain legislative sanction for a new national system in Scotland. It was a specifically Scottish response to the challenges which industrial capitalism posed for elementary schooling.

The universities also retained their autonomy throughout this period, also reformed themselves in response to social change, and also continued to view themselves, and be viewed, as public. Because the Union enshrined their independence in the same way as it maintained the independence of the schools, they were able to continue on the path of modernisation on which they had embarked in the late-seventeenth century, led by Edinburgh but then spreading to Glasgow, to the two universities in Aberdeen, and eventually (after a long delay) to the oldest university in St Andrews. These reforms had laid the basis for specialist appointments and teaching, and for a modernised curriculum that retained breadth of study, placed philosophy at its core, and involved a pedagogy that challenged students by frequent questioning. It was a system well-suited to a poor country with nothing resembling what we would now call secondary education. Boys would go straight from the parish school to the university, and there learn the basics while also engaging in quite deep philosophical speculation. (Girls — although having more or less equal access to the parish schools — had no access to the universities until the end of the nineteenth century.)

On that basis the universities flourished from the mid-eighteenth century onwards, provoking a richness of learning that was Scotland's unique contribution to the Enlightenment. Despite David Hume's notorious difficulties in gaining an academic post — largely because of his atheism — this was a university-based movement, not just in Edinburgh but also in Glasgow and Aberdeen. That educational basis distinguished the Scottish Enlightenment from the Enlightenment elsewhere in Europe. The thinking, teaching and writing that went on then was notable in many respects — for its contribution to a European movement, for its subsequent influence on European and North American thought, and for its capacity to educate to university level a proportionately larger and more socially diverse group of students than any other system in Europe. For the politics of Scottish education, however, two legacies mattered. The leading figures of the Scottish Enlightenment were more engaged in practical social and political activities than were, say, their counterparts in France, especially the second-ranking — but still distinguished — figures such as William Robertson (principal of Edinburgh University) and Lord Kames (judge and agricultural improver) rather than the outstanding luminaries such as Hume, Adam Smith or Adam Ferguson. Indeed, for the scientists such as the geologist James Hutton, the chemist William Cullen and the engineer James Watt, practical applications were the main motive for their work. All these people's writings could be read throughout this highly literate country because of the flourishing system of local printing presses. And that accounts for their second legacy, the political principles which they bequeathed — the belief in civil society (what Hume called 'the middling station') as the basis of effective social regulation. That then became the unofficial guiding philosophy of the self-governing local institutions of Scotland in the nineteenth century.

The universities, like the schools, faced a crisis in the first few decades of the nineteenth century, and they were subject to committees of enquiry which reported in 1826, 1858, 1876 and 1889, with significant reforms in 1858 and 1889. The outcome of these has been hotly debated in the second half of the twentieth century, following George Davie's claims in 1961 that they destroyed the tradition of a 'democratic intellect' — an accessible system of higher learning that counteracted social atomism by breadth of knowledge. But the fullest assessments of the reforms conclude that they emerged from within the system, entrenched the autonomy of the system, partly specialised its curriculum to deal with competition from France, Germany and the revived English universities, and ensured that more students, in practice, actually benefited from the full breadth of curriculum by inducing more to follow the full course that led to graduation. The beginnings of a system of secondary education served the same ends, by allowing the universities to specialise more by gradually withdrawing from the teaching of young students who came straight from the parish schools. By admitting women they broadened their social base substantially — especially to include those able women from middle-class backgrounds who benefited from the new girls' secondary schools in the cities.

Above all politically — as with the schools — the university system continued to be viewed as the property of the nation, much more like the public system of Germany or France than the private colleges of England. This point was made graphically by James Donaldson, principal of the United College in St Andrews. Speaking in 1887, he argued that the Scottish universities were

> not private corporations — they are national seats of learning, existing for the nation, and controlled by the Parliament of the nation. And the Universities have no wish to become independent of the State, or to be removed from the control of the State. (quoted by Anderson, 1983, p. 260)

So, for both schools and universities, the educational reforms of the nineteenth century left a system that was still in fact autonomous and that — more importantly in a political sense — was popularly believed to have responded on its own terms to the need for change. The reforms were instigated and carried through within Scotland's own civic space.

The education system which was maintained and reformed in this period shaped the nation in accordance with a shared social philosophy. Giving it coherence was an attachment to a highly competitive form of social mobility — the belief that the boy of intellectual ability could use education to rise to whatever eminence his capacities allowed. Meritocracy of this type has been called 'contest mobility' by the American sociologist R. H. Turner (1960), in contrast to the 'sponsored mobility' of societies such as England where, traditionally, new members of the elite were recruited at an early age and educated in a wholly separate system of residential private schools. Turner offers the young USA as the paradigm of contest mobility, but it could be argued that such a system favours any peripheral nation trying to open up for its citizens opportunities in metropolitan or imperial centres. A small country like Scotland could not itself give its able young people access to much power, and so sponsored mobility had little appeal. But, by preparing them for independent life elsewhere, it could promote their careers despite that. That is why the various university and secondary-school reforms were indigenous and popular even though propelling many able young people out of the country: they enabled young men (and soon also young women) to compete meritocratically with the products of the English elite schools and universities.

The system shaped Scotland also through its strong attachment to Protestantism and the Union. Especially towards the end of the century, the curriculum was explicitly developed to support Scotland's position in the Empire through what was called the 'concentric' method. That is, children would start learning about their own communities, then widen out to Scotland, Britain and the Empire. Anderson quotes a pamphlet from the Scottish Patriotic Association of Glasgow arguing in 1906 for more Scottish history to be taught because pupils would thereby receive 'a just and honourable conception of the relation of Scotland and England, and would be enabled to understand the beginnings and making of the British Empire' (Anderson, 1996, p. 218). This kind of curriculum is an excellent example of what Graeme Morton calls 'unionist nationalism' — a celebration of Scotland that, because Scotland was conceived of as inescapably British, was also a celebration of the Union (Morton, 1999). That is one reason why the Gaelic and Scots languages were treated with suspicion: they might be useful pedagogically where the children spoke little else, but the purpose was to enable them to become fluent in English. One school inspector opined in 1896 that Gaelic 'will be of much more value and interest when it is dead' (Anderson, 1996, p. 217). These official attitudes were popular among parents: if the guiding principle of the school system was social mobility, then acquiring the language which governed the Empire was much more important than preserving a potentially parochial relic.

Education and the Welfare State: 1872–1965

The renewed system of education after 1872 had to be reformed again as the central state grew in power in the first few decades of the twentieth century. Once more, the pressures for change were international. Education systems across Europe and North America had to respond to the expansion of mass democracy and to the increasingly competitive international economy. The eventual outcome in Scotland (and elsewhere) was a renewal of the sense that the system was autonomous, under the control of national institutions.

The basis of the twentieth-century government of Scottish education was laid with the 1872 Act, which inaugurated a system of locally elected school boards and also established the Scotch Education Department to oversee schools as a whole. This achieved independence from the analogous English board only upon the setting up of the Scottish Office under a Scottish Secretary in 1885. The roughly 900 parish boards were reformed into 38 elected and specialised education committees in 1918, and then absorbed into 35 of the general, elected local councils in 1929. Through two further reforms of local government (in 1975 and 1996), schools have remained under its broad control, governed by 12 Regional and Island authorities from 1975 to 1996, and by 32 councils since then. The Scottish Education Department (as it was renamed in 1918) moved to Edinburgh from London in the 1930s along with most other branches of the Scottish Office (and in 1926 the post of Scottish Secretary was upgraded to Secretary of State). The SED also embraced the schools inspectorate, which had been founded in 1840, and had always been based in Edinburgh.

The SED was a much more assertive department than its counterparts in England, drawing on the faith in state provision which we noted earlier, and was of much higher standing inside the Scottish Office than its counterpart was among the London ministries (McPherson and Raab, 1988, pp. 116–7). It developed its role under its first two secretaries after it became associated with the Scottish Office, Henry Craik (1885–1904) and John Struthers (1904–21). They set the character of the system for the next half century. Craik began the reorganisation of post-primary provision that in the 1930s became the selective system of secondary education, divided by then into senior secondaries which prepared pupils for universities in a five-year or six-year academic course, and three-year junior secondaries which prepared them for work. He also introduced a leaving certificate for the senior secondaries, thus finally making them the normal route into university. Both Craik and Struthers were responsible for consolidating the 1872 system of universal elementary education, gradually winning the struggle to get children to attend right through from ages 5 to 14.

Struthers was responsible for the 1918 Education Act that began the process of making the new secondary system more inclusive. It established the principle of free secondary education for all (but implementing that took much longer), and proposed training for pupils who left school at the minimum leaving age of 14 (an idea which also took many further decades to be developed). The reasons for the delays lie again in the sheer power of the SED: it resisted the most radical interpretations of the Act well into the 1920s. What eventually forced change was pressure from within the system. There was widespread resistance to the rigid separation of children between academic and vocational tracks, and a popular desire to use education to help narrow social divisions. 'The Department', according to one historian, 'discovered that it did not have the power to control events' (Stocks, 1995, p. 59). Secondary education for all (with no more than minimal fees) was in place in Scotland by the late 1930s, and in the selective system the size of the academic sector grew inexorably, so that by the late 1950s it contained some 40% of pupils. This was far beyond the 10% which Craik believed could benefit from full secondary courses.

Equally important in the 1918 Act was its bringing of the Catholic schools into the public sector. For a country and an educational system still dominated by presbyterianism, this was a remarkable entrenchment of pluralism, even if part of the motivation was to control what some (including Struthers) saw as the culturally and ethnically threatening distinctiveness of the Catholic population, with its main roots in Ireland. The Catholic Church accepted the deal because the finances of their schools could not but benefit; there remained very extensive powers for the Church over the curriculum, the recruitment of teachers, and general school management.

Craik also sought to build up an alternative system of higher education, in the largely vocational central institutions and in the specialist colleges of education and the arts. Relationships between the SED and the universities were not good. The SED regarded them as insufferably elitist; the universities regarded the Department (and Scottish politics) as parochial. As the universities moved closer to the state in order to receive public funding, they moved away from the Scottish realm into the ambit of the University Grants Committee in London. The SED insisted on keeping control of teacher training, taking graduates from the universities into courses in the colleges. It also believed that the central institutions could help to regenerate the Scottish economy, following common thinking in the rest of Europe. That division between two branches of higher education remained in place for three quarters of a century, and became itself a source of nationalist campaigning.

Thus the SED led policy-making in Scottish education throughout the twentieth century. Nevertheless, powerful though the SED was, it could not control the processes it supervised, as the example of what happened following the 1918 Act showed. McPherson and Raab argued in their study of the SED's operation in the post-1945 period that it had to recruit allies to ensure that its policies would be put in place. In doing so, it renewed for this state-dominated era the self-governing character that had been inherited from the nineteenth century, finding new ways in which leading educational professionals could maintain their indigenous control. Part of this governing process happened from the centre. In the first half of the century, the SED appointed a series of Advisory Committees filled by educationalists who frequently did not follow the Department's line. Thus it was an Advisory Committee which undermined the SED's restrictive interpretation of the 1918 Act. Most famously of all, it was an Advisory Committee in 1947 which laid the basis for the ending of selection among secondary schools, in a visionary report that sought the most appropriate system of secondary education for a mass democracy (SED, 1947). We return to the outcome of the debates about selective schooling later.

The Advisory Committees were superseded by specialised quangos from the early 1960s onwards, overseeing such matters as examinations, the curriculum, educational technology and community education. Although also appointed by the SED, they, too, tended to bring selected educational professionals into making Scottish educational policy. The SED could not avoid having to bargain with the system. It knew that it could not get its way unless it embedded itself in civil society. Lacking any Scottish source of legitimacy because there was no accountable political process in Scotland — no Scottish parliament — the SED (like the rest of the Scottish Office) could assert its authority only by being seen to emerge from the system and the wider Scottish society. Theoretically, of course, it drew its authority from the UK parliament and the UK state. In practice, what mattered were the roots it assiduously cultivated in Scotland. McPherson and Raab conclude:

> the central roles were not played by national politicians ... [T]he initiatives in policy making lay mainly with the policy community of officials and educationists that linked government with society, and from which Ministers and politicians were for the most part excluded. (McPherson and Raab, 1988, p. 173)

The key SED personnel in this regard were the schools inspectors. Never numbering more than about 100, they did far more than just inspect schools. In effect, they led the making of educational policy, including the appointment of people to the various advisory committees and quangos. Being, technically, professional advisers, they had an authority that derived ultimately from their roots in the educational system itself: their specialist expertise made them permanent fixtures of the policy-making world, unlike ephemeral government ministers in the Scottish Office or the administrative civil servants with their typically British administrative cult of generalism. They sponsored the policy community, building up and maintaining these extensive networks of people who were involved in debates about policy and who took charge of the implementation of policy once it had been made.

The networks themselves were dominated by a few key interests. Most powerful, certainly until the 1970s, were the elected local authorities and their professional advisers, the directors of education. These mattered because they ran the public-sector schools. Until the 1960s, they set the details of large parts of the curriculum (almost all apart from that which was laid down in the syllabuses of the external examinations at ages 16–18), determined teachers' conditions of service, and were responsible in important respects for the structure of schooling — notably the balance locally between junior and senior secondaries. The directors were also respected as authorities on Scottish education, much like the inspectors. So influential were they that the historian of their professional association could claim of the period 1945–75 that

> the unique and central place occupied by the Association in the fashioning of the education structure ... could well support a view that an account of the Association's activities ... would directly suffice as a history of the state provision of public education in Scotland. (Flett, 1989, p. ii)

A second important interest was the teacher trade unions, notably the Educational Institute of Scotland which had about 80% of teachers in membership. Its title indicates its origins in the nineteenth-century campaign to have teachers recognised as a profession, and, despite its trade union activities, it continued to be involved in professional matters. For example, it helped to found the Scottish Council for Research in Education in 1928, and continued itself to commission and debate research, and to send representatives to sit on the various national committees that the inspectorate established. Only with the coming to power of the Conservative government of Margaret Thatcher did its influence wane somewhat, although even then it remained powerful, and it continued to command popular respect (a point to which we return in later Chapters).

A third influential group was the directors of the colleges which trained teachers. Most of them had their own origins in the school system, and so were close to it and to the SED. Others — such as Godfrey Thomson in the 1920s and 1930s — were academics who promoted research as the most sensible basis for planning education. That view was shared by Douglas McIntosh, who moved in 1966 from being director of education in Fife to being principal of Moray House College in Edinburgh. He was an enthusiast for research in both roles. Where the universities were somewhat distant from the rest of Scottish education, and generally keen to see themselves as mainly part of a British realm, the college principals could take on a role of independent intellectual leadership, separate from the SED and yet trusted by it.

From 1965, there was also the General Teaching Council, an element of self-government for the teaching profession as a whole. Elected teachers made up almost a majority of its members (and EIS candidates were an overwhelming majority of them), and the SED officials and assessors were in a minority; the rest were a small number of representatives of other interests such as the colleges and universities. The GTC regulated the profession and approved the training courses that allowed people to gain access to it.

This whole policy community was highly selective socially. The typical route into it — whatever particular segment — was by what McPherson has called the 'Kirriemuir career', after the town in Angus where J. M. Barrie was born (McPherson, 1983). Such places had been served by a single secondary school from at least the late-nineteenth century, and that school would take in almost all the children in the town and round about. Despite heavy meritocratic selection internally, these burgh schools were therefore a site of quite extensive social mixing. They were also academically successful, especially for their brighter children, and sent relatively large proportions to university and the professions. From there, the members of the educational policy community would find teaching posts either back in the same kind of rural school from which they had come, or else in the high-status academic secondaries in the cities. They would step from there into the inspectorate, the colleges or the universities. It was a socially untypical route because it largely by-passed the poverty and relative educational failure of the industrial areas, and yet the memories of social mixing from their own school days gave these educational leaders a sense that Scottish education was egalitarian.

The system which they governed also remained mostly public, just as in the previous century. Until the 1960s, fewer than 1% of the schools were fully independent, and about 3% were 'grant aided' — independent, but in receipt of some public money. Voluntary schools were wholly absent, although that style of provision did gradually grow in the pre-school sector. The system of higher education built up by the SED separate from the universities was as public as anything analogous anywhere else in Europe. The central institutions and colleges were firmly controlled by the Department, and their curriculum was prevented by the inspectorate from straying into the academic territory of the universities (in contrast to the polytechnics in England). So the policy community could readily interpret Scottish education as serving the nation in two senses — through the democratic access it offered to education, and through the public control. A philosophy for all this came from the 1947 Advisory Council report, with its assertion of the essentially public character of schooling. It was the purpose of the state to ensure — especially just two years after the defeat of Fascism — that 'the schools inculcate those virtues without which democracy cannot survive' (SED, 1947, p. 13) as well as preparing young people to contribute to the development of the national economy. The Scottish system of education was 'both democratic and national' (p. 180).

It was not, however, insular, being strongly influenced throughout the century by international currents of educational thought. Because there were more of these than ever before, this sharing tended to reduce the distinctiveness of Scottish education. That has sometimes been described as anglicisation, although it would be more accurate to describe most of it as internationalisation. Nevertheless, the ideas were adapted, often in highly distinctive ways.

An example is in what was made of the idea of selection between the 1920s and the 1960s. Scottish education was enthusiastic about this, as indeed — in the early years — were radical politicians such as the Labour MP Jennie Lee (Lee, 1963). They believed that the scientific measurement of intellectual potential would be much fairer than previous forms of selection, objectively identifying talent in all social classes. This perceived link between selection and social justice explains the character of the Scottish selective system as it evolved. It was more socially open than that in England: a much higher proportion of the population was in selective schools because there were more of them. It was also probably more thoroughly meritocratic, in the sense that advancement beyond the senior years of secondary school depended almost wholly on intellectual attainment, not on social class, although earlier selection points leading to these senior years were as heavily influenced by social class as in any other system in Europe (Gray et al, 1983, p. 217).

A second example is what Scotland made of the parallel growth of child-centred education, the belief that education should start from the child's interests and natural enthusiasms rather than try to impose an externally defined structure of knowledge. These ideas were first used as a critique of the most academic and selective aspects of the system, by radicals such as A. S. Neill in the first few decades of the century, by nationalist sympathisers such as Hugh MacDiarmid (Grieve, 1926), by their successors right through to the 1970s such as R. F. Mackenzie (headteacher of a public sector school in Aberdeen), and by their supporters in the colleges and universities such as William Boyd (founder of the Department of Education in Glasgow University in 1907). The 1947 Advisory Council report brought the ideas into the Scottish mainstream. Again, however, these ideas were adapted, not imported wholesale. Scottish primary schools never developed the same enthusiasm for child-centred methods as those in England or parts of the USA (B. Boyd, 1994). The child-guidance movement which William Boyd pioneered developed in Scotland more as an element of the school's whole ethos than as a separate discipline. And, by the 1970s, the ideas of child-centredness had become closely tied to the ideas of promoting social justice: liberating the individual from social constraints as well as from inappropriate academic ones.

In both these instances — as in many others — the system adapted and so retained its sense of being in charge of its own destiny. It remained self-governing.

Education and the Alien State: 1965–1997

That legacy then began to place the entire system in opposition to the UK state, at the same time as the wider movement for national self government was growing. At first, though, there was no simple connection. The notable feature of education in the 1960s was in fact the strength of Scottish social democracy rather than Scottish nationalism. The Scots took to heart the analysis of social citizenship offered by T. H. Marshall in 1950. Previous, liberal versions of citizenship had put in place civil and political rights — the classic liberal freedoms of speech, belief, assembly, association, and voting. The function of social and Christian democracy, he argued, was to institute a system of social rights without which these liberal freedoms would be unreal for people living in poverty, or with inadequate education to understand how to exercise their freedom. In countries such as Scotland (and also, for example, in Scandinavia) it then became broadly accepted that collective action through the state could mitigate the effects of social obstacles to freedom. In education it was no longer enough just to have open doors — routes for the academically able working-class child into the selective secondary school. These were not enough because many more middle-class than working-class children were able to embark on that route. It was also necessary to compensate for the effects of social disadvantage.

This was another international movement in which Scotland merely shared. In Scotland, as elsewhere — and as Marshall argued — the initial impetus for expanded social rights was mass democracy. Once people had the vote, they could put pressure on the state to serve their interests. To retain its political legitimacy, the state had to respond. It dealt with the pressure in Scotland first by simple expansion, as we have seen — of elementary education at the beginning of the century, of secondary education in the 1920s and 1930s, and of selective secondary education in the 1950s. Each phase of expansion created the social pressures for the next one, and that happened again in the 1960s: the extension of certification by the invention of the Ordinary Grade examination in 1962, taken at age 16, offered the junior secondaries a way of developing more academic courses, and then put pressure for much more schooling beyond age 16 for those who passed these tests.

The cultural and political basis of the new social rights was also much the same in Scotland as elsewhere. There was a powerful Labour movement with a mildly reformist programme. There was a strong attachment to the main elements of liberalism. And there was a legacy of reformist Christianity. Indeed, the 1947 Advisory Council report situated itself firmly in that last current: 'the term "Christian Democracy" ... better perhaps than any other ... summarises the ideals that have governed our thinking about the task of the secondary school' (SED, 1947, p. 6).

But — as earlier — these were international currents that became indigenous. The liberal themes of social mobility that had been current in Scotland in the nineteenth century, and that were still influential on the character of the selective system of secondary schooling in the middle of the twentieth century, could quite readily adapt to ideas of social justice that were associated with the founding and development of the welfare state. Ending selection could be defended as a maintenance of the tradition, even while opponents of selection were also arguing that it was the fairest way of giving educational opportunity to working class children. The Christianity, too, helped to locate the reforming ideas firmly in Scottish history. It may have been called 'Christian Democracy', but — unlike such movements in most other parts of Europe — it was firmly presbyterian, despite the presence of the Catholic schools in the public sector: few

Catholics reached influential positions in the policy community until the 1970s. The whole notion of a system of education that was about public welfare rather than selection also matched well the Scottish idea that the schools were public. A public system ought to serve the public, and that required social justice.

What came out of this further reforming impetus is still with us today, and was the system that was felt to be politically under threat from the governments of Margaret Thatcher and John Major. There were many elements to that. For example, a radical reform of community education was inaugurated by the report of a committee chaired by Kenneth Alexander in 1975 (SED, 1975). It brought together youth work and community development with adult education, and quickly acquired an explicit commitment to social change. It came to be admired internationally for the coherence it brought to community development generally. There were also the first stages of expansion of higher education, providing routes into the professions for many more children of working-class origin, even though the social inequalities in access remained wide.

The most politically potent reform, however, was the system of comprehensive secondary schools. These were officially encouraged from 1965 onwards, although some local authorities (and the Catholic Church) had been experimenting with them since the 1950s. The main feature was the abolition of selection: henceforth, children would attend a full six-year secondary school to which they were assigned according to the neighbourhood in which they lived. This reform proceeded fairly smoothly in Scotland, with much less controversy than in England, and by the mid-1970s almost all public-sector secondaries were non-selective. Most of the grant-aided schools chose to become fully independent, and so about 5% of pupils remained outwith the public system, nearly all in the cities and particularly in Edinburgh, where the proportion was closer to one quarter.

Evaluation of the reform in the 1980s suggested that it had been successful in the terms which had been set. Social class inequalities in examination attainment had fallen while overall levels of examination attainment had risen. So the reform seemed to have benefited working-class pupils without having harmed middle-class ones. Inequalities between Catholic and non-denominational schools had been severely reduced. Before the ending of selection, working-class Catholics were much less likely to be upwardly mobile socially than working-class people of other faiths. For the generation that passed through the fully comprehensive system, the rates of social mobility were identical for Catholics and non-Catholics. The reform also sharply reduced social segregation among schools, although again this was less pronounced in the cities where residential segregation meant that neighbourhood schools were bound to be somewhat segregated. Thus there was more social mixing, one reason why the 1947 Advisory Council believed that a non-selective system was the only type that was consistent with a common citizenship. So generally satisfactory was the new system felt to be that selection for secondary school virtually vanished from serious political debate: by the late 1980s, and right throughout the 1990s, only about one quarter of Scottish adults supported a return to selection, in contrast to over 40% in England (Brown et al, 1999, p. 96).

The comprehensive schools also had other politically potent effects. Although promoting gender equality was not part of the reformers' original intention, the schools in fact had a much greater impact on differences between boys and girls than they did on social-class differences. In part, this was simply another instance of Scotland's sharing in an international experience: the rhetoric of equal rights which social democrats had made popular with the welfare state then became available to further the interests of any social group that did not seem to benefit equally from welfare provision. But it was not only that: the narrowing of gender differences in educational attainment happened earlier in Scotland than elsewhere (in the late 1970s rather than a decade or more later), and so probably did owe something to the early consolidation of the system of common schools. This happened, moreover, at a time when Scotland was closing the few single-sex public schools it had had, as McPherson and Willms (1987) note.

Ending selection had the same kind of consequences for social justice as the waves of expansion in the earlier parts of the century: it forced further changes. The first was a reform of the curriculum and examinations for ages 14–16, attempting to put in place a structure that would cater for the whole age group. Then there was an attempt to establish a common curriculum for ages 5–14 with five common attainment levels, and — in the 1990s — moves towards a common framework of curriculum and assessment for ages beyond 16, under the title of Higher Still. Also due to the same sources was the system of guidance — an attempt to deal with the social and psychological needs of pupils, again in order to reduce barriers to attainment.

What is notable politically about all this is that it was almost all constructed by consensus, and developed and managed by educational professionals. It could thus be felt to be indigenous. Although the initial impetus to comprehensive secondaries came from the Labour government, the momentum was sustained by the local authorities, the inspectorate, and the teachers. The revolution in gender differences owed almost nothing to any kind of official initiative until the late 1980s: it was mostly achieved by enthusiasts in particular schools. Gender equality was then more enthusiastically promoted by the local authorities than by the officials of the SED, and the Conservative ministers barely mentioned it. Although the reforms to examinations and curriculum have been rather more controversial, that has mainly been because of a belief that resources are inadequate to fulfil the hopes of the reforms. There has been nothing like the same resistance to a common curriculum that was evident in England in the 1980s. This was partly because the Scottish curriculum is not statutory (although in practice it is in effect compulsory). Because it was devised by yet more committees of the policy community, now extended to include many more teachers than would have been present up to the 1960s, it could be felt to be a Scottish product.

The reformed system played a crucial role in the politics of the 1980s and 1990s. One reason is that the generation of young people schooled in it had acquired, through that common experience, a firmer attachment to ideas of community and equal rights than their predecessors would have gathered from the selective system. In Scotland as in the rest of Britain, people who have attended school since the comprehensive system was established are much more hostile to selection than their elders (although in Scotland opposition was clear in all age groups). Thus, in the British Election Survey of 1997, among people aged under 35, only 13% favoured a return to selection; the proportion among people older than that was 28%. (In England, the proportions were 25% and 47%.)

This was a diffuse effect on Scottish society as a whole. Education also was important politically for more reasons. The reformed system achieved a unity for essentially defensive reasons, a reaction to the policies of the Thatcher government. There was very widespread suspicion that the Conservatives wanted to privatise the system. Their 1989 Education Act made provision for schools to leave the control of local authorities if a majority of the parents of current pupils were in favour. Eventually only two small schools did so. The reason for the suspicion was the sense of the system as public, and the belief that being public meant being subject to general community control. It is true that there did not seem to be a resistance to parents' involvement in schooling as such. Parents broadly seemed to welcome the opportunity from 1981 to choose a school for their child (something which slowly undermined the integrating effect of comprehensive schools), and they took part in elections to school boards after these were established in 1989. But they wanted that participation to be on their own terms, not the government's. Thus the most notable political impact of the boards was to resist the proposals for compulsory standardised tests in primary schools, something that was felt to be inimical to the child-centred ideas that had become common since the 1960s. Parents' opposition to this was so widespread that it forced the tests to be severely modified, and placed under the control of teachers.

In fact, respect for teachers remained high. This view is found in many countries, where professionals such as doctors and teachers are generally trusted to a much greater extent than politicians. In Scotland, the trust was a source of tension between society and the Conservative government, which tended to denigrate teachers' professionalism: although that rhetoric was more common from Ministers in England, the perception of a government at odds with the teachers spilled over the border. Parents trusted the teachers when they said that imposed external testing would interfere with the less intrusive testing they routinely carried out. They trusted them also when they campaigned against schools' leaving local authority control, and when they argued for better pay in the mid-1980s.

The extent of the abiding trust was measured by a survey shortly after the Conservative government had fallen. It found that 73% of people agreed that teachers 'are in touch with the needs of children', 78% believed them to be hard-working, 76% disagreed that they were overpaid, and 70% believed them to be 'undervalued by society' (ICM for *The Scotsman Education*, 30 September 1998, p. 4). Parents also expressed faith in a public system: 60% of them said they would not send their children to a private school even if they could afford to do so (a similar proportion to that reported by *The Sunday Herald*, 14 February 1999, p. 1). Part of the reason for this trust is probably parents' regular contact with the teachers of their own children, but there is evidence also that people were likely to come into contact with teachers in other settings. Almost all teachers (some 90%) are active locally in civil society beyond education, around half of them taking on positions of leadership in local civic groups (Paterson, 1998a). Not surprisingly, teachers also have views on politics and the constitution which are quite close to those of the population generally (Paterson, 2000a).

Some elements of Conservative policy were accepted — not just the choice of school, but also the common curriculum, the generally consensual approach to the reforms of public examinations and the expansion of higher education in the 1990s. But that is because these were open to interpretations that were quite consistent with the dominant social democratic tenor of Scottish politics and with the belief that education is autonomous. If — as Marshall argued — social rights were the means to realising liberal rights, then measures which increased individual choice were not inimical to Scottish preferences. If the Scottish system was public and served the nation's culture, then a common curriculum need not be an unwelcome intrusion of an alien state: it could include all children in a common national identity. If public examinations were a means to offering meaningful courses to the majority, then reforming them need not be an interference with traditional Scottish practices. And if student demand for higher education was growing because of the very success of comprehensive secondary schooling, then expanding higher education was the best current opportunity to expand social rights, just as expanding secondary education had been in the 1920s. When the Conservative government reformed the system of funding higher education in 1993 to bring the universities as well as the central institutions and colleges under a single Scottish funding body, this was interpreted as the outcome of thirty years of nationalist campaigning against the universities' alleged British elitism. The government themselves helped to reinforce that interpretation by insisting that the change showed how an unreformed Union could still respond to Scottish distinctiveness, and could allow the expansion of higher education to proceed in distinctive ways. It then did: the overall level of participation by young people reached 47% at the end of the 1990s, some 15 points ahead of the rate in England. A much higher proportion of these people attend higher education courses in local further education colleges (about one third, as opposed to one in twelve in England), probably an important means to widening access socially. The FE colleges were also a source of new styles of recruit to university, which attract onto degree courses two thirds of the graduates of higher education diplomas in the colleges.

Conclusion: the Legacy for the Parliament

All this would tend to suggest that the system held firmly to a vision of social democratic rights that successfully resisted New Right intrusions. That is partly true, but matters are not so simple. Whatever the rhetoric, the allegiance to social rights in Scotland has been compromised and partial — not unreal, but also managed by cautious civic institutions. The school structures inherited from the era of comprehensive reform were still, in practice, socially segregated, especially in the cities, and especially there after the effects of parental choice of schooling had been felt. The system of post-school education remained highly segregated: even after the central institutions and colleges were brought into the same funding system as the universities, further-education colleges remained separate and, compared to these higher education institutions, poorly funded. Although social class gaps in access to education had narrowed at all levels, they remained wide. Although the gender differences in attainment had been transformed, women still found access to the higher levels of the professions (including in teaching itself) very difficult. And although Pakistani and Chinese ethnic groups achieved remarkable results in academic attainment, and proceeded to university at twice the rate of their white peers, the system as a whole paid scant attention to racism. Above all politically, despite the rhetoric of a system defending itself against Thatcherism, it remained firmly controlled by the elites — the SED officials, the inspectors, the educational directorates, the school headteachers and college principals. Popular participation — apart from parents' supporting their own children — was in the form of largely passive endorsement of professionals' judgement.

So this was a compromised form of social democracy, not the fulfilment of social rights. It was technocratic management, not popular control. We return in later Chapters to these doubts about the policy community and to the frustration with the only partial fulfilment of the ideals. The main point for the present is that the political legacy for the Parliament is complex. The experience of opposing New Right educational policies contributed directly — as will be discussed in more detail in Chapter 3 — to the main impetus behind the Parliament, to the sense that the Scottish community is taking back into its own hands the key institutions of its national identity. There is, at the same time, a sense that part of the problem is the insufficiently radical character of the education system, governed by unelected elites that the Scottish Parliament has to challenge. And yet these governing groups continue to command respect, partly because of a general scepticism about politics. Scotland shares that with other developed countries, but it may be particularly strong in Scotland precisely because the civic institutions have been felt to have defended Scotland against an alien politics which the country had not chosen. That experience coloured all politics, not just the politics of the right.

The common thread of this Chapter has been the ways in which the education system has been thought to be self-governing. Threats to the perceived autonomy of Scottish education have been repeatedly resented. Memories of previous victories inspire new autonomous efforts. Comprehensive education had that potency in the 1980s, the 1872 system of public schools was cited as a Scottish achievement during the expansion of secondary education in the 1920s, and the 1696 Act (with its predecessors) was the inspiration of the campaign leading to the 1872 reform. What matters for the politics of education is the belief of autonomy, not necessarily its reality, although the belief could not have survived so long if there had not also been some reality to it, as we have seen in this Chapter. Scottish education has been more self-governing than any other aspect of state-directed activity in the period of the Union. Most recently, a sense that the system was under attack and lacked the institutional means to respond has been one of the most potent sources of more general support for a Scottish Parliament. Nevertheless, after three centuries of doing things for itself through the trusted institutions of civil society, Scottish education is bound to find an overtly political self-government an ambiguous opportunity. We return to the problems which this complex legacy will pose for the new Parliament in Chapters 4 and 5. Next, though, we examine in more detail how the various educational traditions and recent debates we have looked at here entered into the campaigning that achieved that Parliament.

Chapter 3

EDUCATION AND THE POLITICS OF SELF-GOVERNMENT

No significant educational interests opposed the setting up of a Parliament during the referendum of 1997. In contrast to the disagreements during the referendum on a Scottish Assembly in 1979, the teachers' and student leaders, the college and university principals, the elected local authorities and almost all the regular writers on educational politics — educationalists, journalists, academics — all seemed to agree that education could benefit from home rule. The Educational Institute of Scotland contributed money and campaigning advice to the main organisation which advocated a vote in favour of a Parliament, and sponsored conferences and publications examining the implications it would have. It would certainly not have been possible to say the same of Scottish educationalists in 1997 as John Smith did of Scottish professionals in 1979 that

> I am afraid there is something about the Scottish professional middle-class — something which makes them take to their heels and run, when a real test arrives. They deserted in droves, after having been — equally characteristically — only too happy to toy with the notion of self-government, after being agreeably intrigued with it for some time. (Smith, 1981, p. 50)

The new unanimity was the culmination of two decades during which the protection and development of Scottish education had come to be aligned with a Parliament. We have already noted part of the explanation of that, in Chapter 2, where we saw the broad political reasons why educationalists could no longer be as enamoured of the Scottish Office as Smith found them to be in the 1970s — the tensions between a Scottish social democracy that was felt to be autonomous and a Conservative government that was determined to reform it. This Chapter examines the consequences of that political experience for the debates about self-government: on what grounds were educationalists persuaded to abandon their previous caution?

A Scottish Parliament and Educational Democracy

The most common and most obviously appealing arguments were that a Scottish Parliament could subject the education system to proper democratic scrutiny, and could make better educational policy.

The democratic point simply applies to education a more general one about the Scottish Office as a whole. Ever since John P. Mackintosh's campaigning in the 1960s, it has been argued that running a nation by means of a separate bureaucracy was indefensible in a democracy. If the UK was a single realm, then there was no need for a separate branch of administration in Scotland. If it was not — as was very widely agreed — then a separate administration was not enough. This kind of democratic argument appealed particularly to social democrats who were suspicious of campaigning based on nationalism. It is the point of view most associated with Labour and Liberal leaders such as John Smith, Donald Dewar and David Steel, although it has also formed part of the case made by the social democratic wing of the SNP.

The democratic argument in its most general form noted the distance between the civil servants of the Scottish Office and the parliament in London that was meant to oversee them. Most non-Scottish MPs are uninterested in Scottish matters, most Scottish MPs spend most of their time in London, and Scottish Office ministers had to manage too many separate interests to be fully in command of their briefs. John Smith again put this well, drawing on his experience as Parliamentary Private Secretary to the Scottish Secretary of State, Willie Ross, in the 1960s. The minister, he said,

> is down at Westminster from Monday to Thursday; the civil servants in Edinburgh have some diversion ready and waiting for you on the Friday; and the result is you have no real access to or grip upon the senior administration. (Smith, 1981, p. 46)

He described the resulting system as 'mandarin government'.

There have also been specifically educational aspects to this point. Scottish Office ministers of education had to contend not only with the administrative civil servants but also with the powerful inspectors. As we saw in Chapter 2, these have commanded authority through their educational expertise. They are also able to assert their independence of the political process on the grounds of being professional advisers, appointed by the Crown, rather than servants of the politicians or even of the administrative machine. Their power also rests on their access to detailed knowledge of the education system — their inspections of schools, their commissioning and presentation of research, and their oversight of the educational quangos, appointments to which they, in effect, control. The inspectors have a strong influence over how teachers and community education staff are initially trained in the colleges of education (now mostly the university faculties of education), and so can shape the outlook of the country's educators.

Added to that has been a more sociological point about the social unrepresentativeness of both the administrators and the inspectors. In McPherson's terms outlined in Chapter 2, they have reached their positions through the Kirriemuir career: socially, they are highly untypical of the students or indeed teachers over whom they preside. That selectivity may have declined somewhat in the more meritocratic spirit that accompanied the full development of comprehensive education, but a large degree of social remoteness remains at the most senior levels (which — despite Conservative reforms — continue to recruit mainly from people who entered the governing system in the 1970s and earlier). For example, of the four heads of the inspectorate and the four secretaries of the SED who held office between the mid-1970s (the end of the period covered by McPherson) and the setting up of the Parliament in 1999, all but one attended a secondary school that had been selective before 1965, seven studied at an ancient Scottish university, and one went to a college in Oxford. They were also all men. None of them — apart perhaps from the head of the inspectorate since 1996, Douglas Osler — was at all known to most teachers, far less to the general public. Compared to elected politicians — or indeed to classroom teachers — this governing group had much less routine contact with people who were distant from power.

Beyond the SED itself were the quangos, easy targets for the accusation of lack of democracy because they were even more remote from parliamentary scrutiny than the civil servants and inspectors in the Department. For example, during the six years between the founding of the Scottish Higher Education Funding Council and its transfer to the new Scottish Parliament in 1999, its chief executive had never been asked to appear before a committee of the House of Commons in London. The term 'quango' can be a matter of arcane dispute among political scientists as to its exact meaning; the official title is Non-Departmental Public Body. In popular usage, it signifies any committee outwith the structures of elected local councils or parliament which has executive power and control over a budget. It thus refers to the large number of bodies which regulate the technical content of Scottish education — setting the examinations, determining the curriculum, funding higher and further education, or advising on the development of community education.

Attitudes to such bodies — and indeed to the inspectors — are probably more ambivalent in Scotland than in, say, Wales, where quangos were largely a product of the 1970s and 1980s, and so were readily associated with Conservative control of a country that had not endorsed Conservative policies. That view was certainly present in Scotland: there was a feeling in the late 1980s that the Conservatives were appointing their own supporters to the committees, in contrast to the view which prevailed up to the 1960s that the criterion for appointment should be non-partisan expertise. There was a feeling also that inspectors were serving their political masters too slavishly: for example, when the Conservative government was trying to enforce its original scheme of national testing on primary schools in 1989-90, a lot of the proselytising was done by Douglas Osler (then a senior member of the inspectorate, but not yet its head). On the other hand, because many of these modern quangos had their origins in committees that went right back to the nineteenth century — for example, the examination board — there was also a sense in Scotland that they were part of civil society, not just of the state. This capacity of Scottish quangos to distance themselves somewhat from the Conservative government was probably aided by the inspectors themselves, who almost certainly were able to lobby discretely against government policies. Sometimes the inspectors had significant and public achievements in that regard: for example, in 1984, they successfully resisted the encroachments of the Sheffield-based Manpower Services Commission into Scottish vocational education, allowing time for a separate Scottish system of vocational qualifications to be set up.

Nevertheless, even this capacity of the inspectors and the quangos to modify government policy has been used as an argument for a Scottish Parliament, because the processes are so opaque. No-one really knew systematically what subtle adjustments were made by officials to ministerial intentions, and this secrecy has been felt to be inconsistent with democracy. Even if the quangos are open to persuasion by interest groups, that — it can be alleged — is still not fully democratic because it gives too much power to educational insiders, and not enough to the elected representatives of the people. Only an open process of parliamentary scrutiny, it is believed, can change that.

Beyond these specific points about democratising the policy communities, there has been a sense that Scottish education has somehow embodied Scottish democracy throughout the period of the Union, and so that a Parliament could serve Scottish education because it would entrench that democracy more firmly. The nineteenth-century belief that education was socially open, and its twentieth-century translation into the language of social democracy, has created an impression of a system that serves the people, as we saw in Chapter 2. The most recent instance of this is the popularity of comprehensive secondary education — the belief that it most accurately embodies a Scottish tradition of democratic education. Indeed, it is this much more recent version of the argument that has made it congenial to the social democratic proponents of a Parliament. Talk of ancient democracy may appeal to nationalists, but not to modernisers. But when the educational democracy is in the tangible form of a recent, successful innovation — one, moreover, inspired by the Labour party — then it seems safe for current use. It is a way, moreover, of linking the democratic case to the better-government one (to which we turn shortly): for a polity attached to ideas of social justice, promoting wide access to education is both democratic and also good policy.

A similar link to democracy has been claimed for the tradition of local self-government, from the parishes of the eighteenth and nineteenth centuries to the still powerful local authorities today. If education has been traditionally democratic, the argument goes, then there is no better way to realise that democracy fully than to place it under the supervision of a democratic national parliament. Malcolm Rifkind, when Secretary of State, appreciated the potency of this argument for the debate about home rule when he claimed that the Conservatives' scheme for schools to leave local authority control was actually more in keeping than were local authorities with the tradition of local self government (Paterson, 1994, p. 2). But Rifkind and his allies lost that argument, and in fact educational opposition to the Tories became another reason to claim that the system is distinctively democratic. It sided with the people in their darkest hour.

A Scottish Parliament and Better Policy

These democratic arguments for a Parliament have appealed to constitutional reformers and to people who see the education system close up. Probably more effective at a popular level, however, has been the simple claim that a Scottish Parliament would make better policy than Westminster. We noted (in Chapter 1) the long tradition of such campaigning and the findings of the 1997 Scottish Election Survey that a very clear majority of people believed that a Parliament would improve education. That is probably why a large majority also believe that a Scottish Parliament is the best agency to be charged with governing education (64%, as opposed to 5% for Westminster and 25% for local councils, according to a poll conducted by ICM for *The Scotsman Education*, 30 September 1998, p. 4).

Various reasons have been advanced to support this belief. One is primarily about perceptions. It could be said that anything with the word 'Scottish' in its title tends to command more popular support than anything based in London. That is not a trite point: if it is true, then there might be a greater general willingness to work constructively with the Scottish Parliament than with the unreformed constitution, and so the new Parliament could have an impact more because of how teachers and others react to it than necessarily because it makes better laws.

The claim of better policy has also been linked to the claims about more democracy. This has been around for a long time too. John P. Mackintosh argued in 1975 that:

> Scottish education is not only headless but its control is fragmented between the staff of [teacher] Training Colleges, Scottish Office civil servants, the Inspectorate and some leading teachers. But there is no political point of decision-making which would draw these groups together and make them face the options in a Scottish context. As a result it is not clear who does control Scottish education, the major issues go by default, there is no sense of leadership and in practice the only thing left is to tag along behind the decisions of the English Department. (Mackintosh, 1975, p. 6)

It has been argued that a Scottish Parliament could give more time to debating education, and that it could reach conclusions through rationally weighing evidence rather than from political prejudice. Inspired by this sentiment, the standing orders of the new Parliament have been based on the 1998 report of the Consultative Steering Group. The intention is that legislation would be more carefully drafted than at Westminster, and be subject to a much lengthier and more thorough process of consultation. Running through this is a sense that politicians cannot be trusted to make good laws at all, a rather paradoxical point when the institution in question is an elected Parliament and the people with whom they are liable to consult are the very professionals whose activities are supposed to be coming under greater democratic scrutiny. We return to such apparent contradictions in Chapter 5.

It has also been argued that widespread consultation will ensure that policy is better implemented because teachers and others will be less sceptical of it. The role of implementation in determining the quality of a policy is familiar in Scotland: it was teachers who prevented the most controversial aspects of Conservative policy — such as national testing — from being implemented at all, and it was teachers who shaped those policies that have made any difference to education, such as the curriculum for ages 5–14, the examinations at age 16, and the new Higher Still examinations in the years following that. One study of teachers' attitudes to the 5–14 curriculum concluded that

> while politicians and managers of education systems have the power to offer rewards and to impose sanctions to encourage teachers to innovate, it is teachers themselves who ultimately decide whether or not an innovation will be implemented in the classroom. (Brown and McIntyre, 1993, p. 117)

It has been argued further that a Scottish Parliament would make policy that would better suit the diverse needs of different parts of Scotland because it would be more responsive to local concerns. The same devolving impulse that set it up would decentralise decision making beyond Edinburgh, if only because the dominant political parties are aware that rural Scotland is sceptical of political influence from the central belt. Scottish education has been accustomed to quite tight central control since 1872, and this has come to be seen as yet another failure of the Union. It has been suggested that the standardisation has been artificial. It has been partly defensive, the necessary unity that would allow SED officials to extract resources from the Treasury in London. And it has been based on the social homogeneity of the elites which run the system, the beneficiaries of the Kirriemuir career who then impose their partial understandings of the system on everyone else.

Nevertheless, above all — as we saw in the last quotation from Mackintosh — a Parliament has been thought to be good for education because it would re-unite it. The Union, it is claimed, has fragmented a once-coherent system, offering a route from parish school through university into the professions. That old unity was not retained during the necessary process of modernisation in response to social and economic change. Secondaries, it is alleged, do not relate smoothly to primaries, vocational courses are still regarded as inferior to academic ones, higher education is divided between colleges and universities, universities themselves are too concerned with entrants from England to meet the needs of people with Scottish qualifications, and universities still suffer from their 75 years of governance from London. Such aspirations after national unity are recognisable features of a broadly nationalist movement (however much social democrats might decry that label). It is believed that coherence is good. Romantic that may seem, but it could also become true if it inspired a unity of national purpose, a belief among most of the partners to education that they were now working towards a common end.

A Scottish Parliament and Public Education

The other educational arguments for a Scottish Parliament have tended to be of that last type — as with the putative ancient tradition of educational democracy, less concrete, more to do with fulfilling aspirations than with the details of policy or the policy process. For example, a Parliament has been felt to be the best way of maintaining the public character of Scottish education. That genuinely long-standing tradition was maintained throughout the years of the Conservative government, again as we saw in Chapter 2. Even within the unreformed Union, and well before the controversial intrusions of the Conservatives, there was a stronger line of specifically Scottish legislation on education than on any other topic of social policy, and this practice tends to confirm that the whole system is public property. It has then been argued that a public system is best safeguarded by that supreme public body, a national Parliament, aggregating particular concerns. Such claims have become familiar in many nations since Hegel first made them systematic in the early nineteenth century. Only a strong national state, he claimed, could ensure that all the particular interpretations of the public interest be reconciled. That view became a standard part of nationalism, and, in a different form, of socialism.

One prominent example in Scotland recently has been campaigning to repatriate the universities and to make them more Scottish. University politics were central to many nationalist movements in nineteenth and early twentieth-century Europe, as politicians sought to give their countries' culture the recognition of academic scholarship, and also sought to retain their socially mobile young people by educating them for professions locally. Ireland and Wales shared in this general tendency, but Scotland did not, because it already had its own autonomous universities. Nevertheless, in the different world of the 1970s and 1980s, memories of these earlier nationalisms elsewhere helped to make rescuing the public character of the Scottish universities a recurrent feature of campaigning for a Parliament. George Davie's thesis that the universities had betrayed their Scottish roots gained currency, helped by the hostility of university leaderships to a Scottish Assembly in the 1970s. Although the Conservatives defused the issue politically in 1993 when they placed the universities under a Scottish funding body, it has not gone away. Its most recent manifestation has been the fervour of the campaigning to abolish the tuition fees that were introduced throughout the UK by the Labour government in 1997: universities that were public property would — it is claimed — not charge for their services. That focus also conveniently linked the public character of the universities to the wider question of social justice, and so also to the belief that the Parliament could make better policy.

Even more intangible is the claim that a Parliament could renew the sense of Scottish community, an argument that has been particularly popular on the left of the SNP where it has seemed to offer an acceptable way to avoid the more xenophobic types of nationalism through keeping a link to decentralising versions of socialism. People such as Isobel Lindsay, Stephen Maxwell and Neil MacCormick were making such points long before their party was taken over by social democrats such as Alex Salmond (and even longer before the Labour Party gained its current Blairite enthusiasm for 'community'). For example, in 1976, Lindsay went back beyond the statism of the dominant currents in British social democracy to the community socialism of G.D.H. Cole, and argued that

> a political unit can provide a goal for collective achievement, something which has greater continuity through time than any individual. It may reinforce community at the same time as requiring community to provide its coherence. (Lindsay, 1976, p. 26)

This view of community found the idea of community schools and community education congenial, and so fitted well with some of the most popular features of Scottish education in the 1980s and 1990s.

The final way in which a Parliament was claimed to be able to renew the public character of education had to do with what Habermas has called 'communicative space'. It has been argued that, if democracy is to be renewed, then it has to be on the basis of getting people to talk to each other, as it were — overcoming privatism and social isolation by forging communities based on rich social communication. Stewart Ranson, writing about such aspirations in England, has proposed that this is the link between democratic and educational renewal: the more you educate people in common, the more they form a proper community, and so the more able will they be to hold their elected institutions to account (Ranson, 1994). Several such writers explicitly point to the Enlightenment period in Scotland in the eighteenth century as an example of this social dialogue in practice, and as providing the conceptual tools for renewing it. In that sense, an education system that serves a community (rather than, say, only the career aspirations of individuals) would be helping to renew democracy, and would acquire from that role a proper sense of its own public function. Enthusiasts for this kind of idea in Scotland have also tended to point to the opportunities which new electronic technologies offer for renewing both education and democracy. Allowing technological access to the new Parliament — and encouraging educators to use it — has become a standard part of the claim that it can renew democracy.

A Scottish Parliament and Citizenship Education

The proposal that a Scottish Parliament can renew citizenship is part of the wider campaigning for constitutional reform throughout the UK, notably by the organisation Charter 88 and in the report on citizenship education in England from a committee chaired by Bernard Crick. The debate about citizenship on the reforming left arose in the 1980s as a response to the enthusiasm for individual freedom by thinkers on the New Right. It was believed that this rightist version was too individualistic, and paid insufficient regard to the intrinsically social character of citizenship. A citizen is a citizen *of* somewhere. There was a ready-made philosophy of this available by going back to T. H. Marshall (whose ideas were summarised in Chapter 2): just like rights, citizenship was not only individual but also social. In the words of Stuart Hall and David Held,

> citizenship rights can be thought of as a measure of the autonomy an individual citizen enjoys as a result of his or her status as a 'free and equal' member of a society. (Hall and Held, 1989, p. 177)

Emphasising citizenship would also offer the left a way of responding to the New Right's claim that the social democratic state was too powerful, and sapped individual responsibility and initiative. Vigilant citizens could hold the state to account.

These ideas have Scottish roots too. The point about active citizens embedded in a social context owes its origins to Adam Ferguson, and was part of the guiding philosophy of Scottish local democracy in the nineteenth century. Yet they were not a strong part of recent Scottish social democratic thinking (except, again, in the social democratic wing of the SNP). Debates around the Constitutional Convention began to provoke a new attention to them, partly because of the membership of many Labour and Liberal Democrat activists in Charter 88. By the late 1990s, there seemed to be general agreement that education had a role in underpinning the Scottish Parliament through offering young people and adults a chance to learn how to be active, responsible and critical citizens. This view gained official endorsement in the report of the Consultative Steering Group, and in a report on Scottish culture in the school curriculum from the Scottish Consultative Council on the Curriculum:

> The current constitutional reform process and the establishment of a Parliament in Edinburgh should be seen as golden opportunities to create a new political culture in Scotland. The curriculum has a key role to play in this task. The new culture must be one of inclusiveness which seeks to inform and engage young people by developing their understanding of political processes and issues. (SCCC, 1999, p. 8)

The idea of creating a new political culture had come originally from the women's movement, campaigning for an equal number of women and men in the Parliament partly on the grounds that women tend to civilise debate. It became common to cite Scandinavian evidence that, beyond about 30% women, the style of politics does change.

It has also been argued that citizenship itself is impossible to define, being more about capacities than about particular forms of achievement or particular knowledge. There is an analogy with thinking skills, to which educators have been turning for means by which students can learn effectively: the attention is to the skills, but they are best exercised in a specific domain of knowledge. In developing these capacities, there are three aspects to the role of education in preparing people to be citizens, each of which has found resonance in the debate about a Scottish Parliament (Avis et al, 1996).

The first is the simple provision of facts about democracy, as instanced in the education service set up before the Parliament was even elected and the quick introduction of a topic on the Parliament into the syllabus of the relevant school subject (Modern Studies).

The second is the view that citizenship is best promoted by a good general education, because that reduces the barriers to social dialogue; this relates to the concept of creating 'communicative space'. General education was claimed to be part of the main Scottish tradition, and entered nationalist debate in the 1980s through the disputes about university culture that were inspired by George Davie's writing. Early specialism divides people, it is claimed, and consequently many decisions of government are abdicated to committees of experts.

The third way in which it is believed that education can help to develop citizenship is through giving people an experience of autonomy in the process of learning. One writer has called it 'education for studentship', and by studentship he means the 'capacity for independent study and for recognising the problematic nature of knowledge' (Bloomer, 1996, p. 140). This has entered recent Scottish educational practice through the child-centred ideas of the 1960s, which — as we saw in Chapter 2 — became one motive for education's resistance to the policies of the Conservative government (notably the proposal to introduce standardised testing in primary schools).

All these ideas on citizenship have tended to create the sense that education is part of a wider project to renew Scottish democracy. Renewing citizenship, especially through education, seems nicely rational, far removed from nationalist passion. Of course, it also links with nationalism whenever attention is directed to the character of the community to which the citizen is meant to have allegiance. But that issue, too, has been avoided, even by nationalists, keen to avoid the concomitant question of who should be excluded. Nationalist interest in education and citizenship would prefer to point to the experience of Catalonia, where the education system has been used to socialise large numbers of immigrants into the society that has been ruled over by a powerful home rule Parliament since 1980. Then, in Scotland, that theme of social inclusion links neatly back to the claims of social justice that have shaped the education system since the 1960s.

A Scottish Parliament and Scottish Culture

Debates about citizenship are a particular instance of more general debates about culture, the final way in which a Scottish Parliament has been felt to be relevant to education. This returns us to some of the classic nationalist arguments linking education and self-government — the nineteenth-century programmes which viewed education as the means to national renewal. Some examples from Ireland were cited in Chapter 1. Here is another example, from Norway, illustrating both the role which nationalists saw for the state in this process and also the extent to which such ideas pervaded social democratic parties too: the author was the Labour Foreign Minister, writing in 1937.

> It is the great goal of all labour movements ... to build up a national folk culture, and it is therefore natural that the working class now joins with the farmers' movement in support of the language programme which will lead to complete national unification. (Derry, 1973, pp. 334–5)

Similar views on the role of education in creating national unity can be found from Scottish socialists speaking in the same period. Thus James Maxton, in the 1925 speech to the EIS which was cited in Chapter 1, tied his support for a Scottish Parliament's control of education to the renewing of national culture:

> the people of Scotland and the teachers of Scotland [have] a duty ... to discover definite national ideals — ideals which will not be contradictory of national traditions, which would not clash with national characteristics, but which would be complementary to these and inspire them with new life and vigour. (Maxton, 1926, pp. 38–9)

One reason why cultural renewal has featured in the Scottish debates has simply been a memory of this history. It is expected of nationalist movements that they should revive the national culture. For example, in 1988 the Claim of Right which inspired the Constitutional Convention had this to say:

> The twentieth century ... has been a period of extraordinary fertility in all fields of the Scottish arts. ... We think it no accident that this trend has accompanied an increasingly vigorous demand for a Scottish say in Scotland's government. (Edwards, 1989, p. 14)

The fact that a lot of this activity has been financed by the organs of the UK state — the Arts Council as well as the education system — is either not remarked on, or else subsumed into the argument that these are, anyway, a part of Scottish civil society, not really the state at all.

That is one source of thinking. It has now produced a widespread assumption that the new Parliament can help to renew culture, and that the education system can be part of that. The review of Scottish culture in the curriculum mentioned earlier is a good (and officially sanctioned) instance:

> The school curriculum is not only a manifestation of … transmitted culture, but also plays a key role in supporting and sustaining it.
> The question of national identity seems likely to continue to be the subject of public discussion. … Inevitably there are implications for school education. (SCCC, 1999, p. 6 and p. 3)

At the same time, however, the character of that culture has been felt to be changing, and there has been much more attention to pluralism than there would have been in earlier nationalist movements. This is not straight-forwardly a matter of contrasting the nationalist arguments for a parliament with the social democratic ones. Both movements used to celebrate homogeneity (or unity), and both now celebrate Scotland's multiculturalism. Here are four statements from the manifestos that were published during the campaigns for the Scottish Parliament elections in May 1999, together illustrating the coexistence of the view that the Parliament could promote Scottish culture and also protect cultural diversity; out of context, only rhetorical nuances separate the nationalist and social democratic versions:

> The arts and culture have a central role in shaping a new sense of community and civic pride in the new Scotland.
> The Scottish Parliament should mark a new phase in the confidence in our cultural life.
> Multicultural education should be mainstreamed into all areas of the curriculum.
> We will celebrate the diversity of Scotland's people.

(In fact, the first and the last come from the Scottish Labour Party (1999, p. 19 and p. 17), the middle two from the SNP (1999, p. 27 and p. 22).)

This multiculturalism also features in the official review of Scottish culture in the curriculum:

> Scotland is a pluralist country which has, over the years, absorbed and been enriched by incomers from other parts of the world.
> Debates [about Scottish identity] should give due attention to the history and culture of Scotland, while recognising the richness and diversity of the Scottish experience and its location in a wider context. (SCCC, 1999, p. 5 and p. 16)

Tensions between the Scottish Parliament and Education

All these arguments linking education to the campaigning for a Parliament would tend to suggest that the two would happily coexist. The very force of some of them — and certainly their popularity — will create strong pressures for that to happen, as we will see more fully in Chapter 5. But there are gaps in many of them. That is not surprising: political arguments are not meant to be intellectually watertight. The gaps can, however, indicate the ways in which the campaign for a Parliament has not been entirely at ease with promoting the education system as it is, and can therefore suggest (in Chapter 5 too) the potential for acute tension between the Parliament and education.

Lying behind the complicating themes is a contradiction that is inherent in all nationalist movements in conditions other than those of extreme oppression (which Scotland has never faced since 1707). If the Union has been so heinous (or the Union in the form it took from 1707 to 1999), then how can Scottish education have anything to celebrate? Alternatively, if Scottish education does have a lot to be proud of, then perhaps the educational case for a Parliament was not so clear cut after all?

The most notable general version of this has long been associated with the writing of Tom Nairn. The core of his argument has been to point to what he believes to be the thoroughly unsatisfactory character of the civic government which Scotland has had within the Union:

> Institutional identity seems to me broadly the same as managerial identity or, less flatteringly, ... 'bureaucratic identity'. The self-management of civil society historically found in Scotland implied a class which administers and regulates rather than 'rules' in the more ordinary sense of political government or direction.
>
> Bureaucracy had an egalitarian side to it and within limits was a social leveller, especially in contrast to aristocracy or older, class-bound societies. But that should not be confused with democracy. (Nairn, 1997, p. 205)

Whatever the merits of Nairn's argument, the point for the present analysis is its wide influence over not only some segments of the SNP but also the nationalist-inclined parts of the Labour and Liberal Democrat parties, and even some parts of the radical right among the Conservatives. It is connected also with the argument which links education to home rule through the democratising case: Nairn's points are, in some respects, simply a much more radical version of that.

One manifestation in education is frustration with the social conservatism of the governing groups. The most noted exponent of this view is Walter Humes, whose 1986 book on the leadership class in Scottish education provides the alternative way of describing the policy community towards which McPherson and Raab are fairly sympathetic. Humes wrote:

> Those qualities with which members of [the leadership class] have been associated — bureaucratic expansionism, professional protectionism and ideological deception — hardly amount to a vote of confidence in their collective achievements. (Humes, 1986, p. 201)

He concludes that the main cultural effect was to promote 'unreflective conformity' (p. 208). He also follows Nairn in arguing that Scottish civic life is deeply apolitical. Any move for radical reform would

> be resisted by those who have lived by the myth that Scottish education is autonomous and who argue that it ought not to embroil itself in the contaminating world of politics. (p. 204)

There is indeed some evidence that school teachers in general are not particularly radical, holding views on civil liberties that are just as conservative as those of the Scottish population as a whole (Paterson, 2000a).

Humes adds that criticism of the leadership class is difficult to find, because people who might be critical either emigrate or are in thrall to its capacity to control the avenues of career progression. Nevertheless, partly inspired by Humes's (and Nairn's) critique, similar views have been expressed rather more frequently since then. One contributor to a book in 1990 on girls' and women's experience of the governing system said this:

> Women in Scottish education are disadvantaged by structural and systematic exclusion. ... Appointments are almost invariably made by men who appoint in their own image. ... The products of [the] self-replicating working groups [on curriculum reform] are likely to remain dominated by a particular male perception of the world, characterised at best by female stereotypes and at worst by disregard and contempt for women's history. (Hills, 1990, p. 162 and p. 164)

That kind of view lay behind a lot of the feminist campaigning to achieve a high proportion of women in the Scottish Parliament.

Part of the conservatism has been a solid unionism. The leadership class may have governed a semi-independent social policy, but for that very reason they were highly reluctant to push for any greater autonomy. Nairn argues that the managerial ethos has induced such a pathological caution in the Scottish institutional middle class that it is unable to do anything radical at all, and even repeatedly unable to make up its mind on whether there ought to be a Scottish Parliament. In 1991, he wrote:

> Its dilemma of the present day — under the Thatcher onslaught — is ... whether or not the moment has come for it to rule, rather than merely administer. (reprinted in Nairn, 1997, p. 187)

This argument is reminiscent of John Smith's in 1981, explaining the Scottish middle class's reluctance to support home rule.

From that there is an equally frustrating 'display nationalism', confined to football terraces, rugby stadiums and the rituals of Burns night and Hogmanay, purporting to be radical but actually deeply conservative and hierarchical. Writers on education have made many similar comments, notably in the highly influential collection of essays which Walter Humes and Hamish Paterson edited in 1983, *Scottish Culture and Scottish Education*. Its unrelenting picture of dull conformity resonated in T.C. Smout's characterisations of the system throughout the last two centuries:

> It is in the history of the school more than in any other aspect of recent social history that the key lies to some of the more depressing aspects of modern Scotland. If there are in this country too many people who fear what is new, believe the difficult to be impossible, draw back from responsibility, and afford established authority and tradition an exaggerated respect, we can reasonably look for an explanation in the institutions that moulded them. (Smout, 1986, pp. 209–30)

Hamish Paterson, similarly, argues that the national culture created by the education system has been divisive and hypocritical:

> If Scottish schools were ever democratic, they were democratic in a particular way which emphasised social division, competitive liberalism and individual achievement at the expense of others. (H. Paterson, 1983, p. 205)

Humes (1986) describes the resulting nationalism as 'complacent' (p. 97), feeding 'the susceptibility of many Scots to a form of nationalistic flattery that is designed to ensure docility through reassurance' (p. 9).

Such critics have tended to assume rather than demonstrate that the governing system of Scottish education is homogeneous. Whatever may be the truth of the claim that teachers have tended to be conservative and conformist, on the whole that would matter less for policy than the alleged conservatism of the people with substantial amounts of power: the schools inspectors, the senior educational officials in the Scottish Office, the directors of education, the conveners and chief executives of the main educational quangos, the university principals, the leaders of the teachers' trade unions, and a few local and national politicians.

We return in Chapter 5 to consider the heterogenity of the governing system, which may matter rather more now that a Parliament has been established (with all the multiple channels of political expression which a more public form of politics allows). For the moment, however, the main point is about the inferences which have been drawn from the critique of elite caution. One conclusion that is especially common among supporters of a Scottish Parliament has been the urge to use the Parliament to make much more of a reality of the aspiration to social justice that underlay the welfare state and comprehensive education. It has been argued by many radical groups that the Parliament should promote equal opportunities more stringently than the conservative Union has done. It is claimed that the opportunities which have been created by mass secondary and higher education are based on meritocratic competititon rather than on equality of average outcomes between social groups. For example, comprehensive secondary education removed the most salient institutional barrier to working-class achievement, but that is regarded by radicals as insufficient.

The attempts by subsequent policy to compensate for the continuing educational disadvantage of people living in relative poverty are regarded as merely tinkering with the educational effects of structural disadvantage, and some further changes — such as ending maintenance grants for students attending full-time courses in higher education — are seen as actually working in the opposite direction. So it has been proposed that the Parliament should renew comprehensive education, not undermine it as the Conservatives were alleged to have done, and as the Labour government is alleged to be doing in England. It is argued, too, that child-centred ideas should be renovated, not rejected, by adopting international ideas on emotional intelligence, reducing the extent to which education takes place in invidiously institutional settings such as schools, and embracing the most creative potential of the Internet. None of this, it is claimed, would be likely to come from the cautious reformism of the old Union. Nationalist radicals also argue that little of it would come from the cautious Labour Party.

On the other hand, since the leadership class was also quite attached to a very gradualist social democracy, there is some similarity between these views from the nationalist left and the views of radical right-wingers such as Michael Forsyth when he was Education Minister in the Scottish Office between 1987 and 1992, and then Secretary of State from 1995 to 1997. He repeatedly made clear his contempt for what he saw as the complacent and woolly liberalism of the influential groups in Scottish education. That view is now dominant in the pages of *The Scotsman* newspaper, campaigning for educational reform on the grounds that standards throughout Scottish education are poor, that the whole comprehensive idea was misguided, that teachers, local authorities and inspectors are complacent, and that only radical political action can solve these problems. Also from the right is the analysis of the historian Michael Fry, who reaches conclusions that are similar to Nairn's and Humes's on the deleterious effect of Scotland's civic politics:

> The Scots … have … forgone some of the principal benefits of attachment to the English parliamentary constitution. Partisan spirit and alternation in power do … foster constant critical scrutiny, diversity of opinion and efforts at reform. In Scotland, by contrast, government is conducted behind closed doors. Debates are predictable and lifeless. … The consensus, however well-meaning, inevitably turns inert, inflexible and hostile to novelty. (Fry, 1987, pp. 255–6)

We look at the right-wing critique of Scottish education in some more detail in Chapter 4. It played little part in the campaigning for a Parliament, although Fry has also consistently belonged to that small band of Conservatives who remained committed to one. Now that the Parliament is in place, his line is the obvious unifying philosophy for the party. The new Scottish right — for example, the Conservative education spokesman Brian Monteith — has accepted the Parliament they once opposed, and see it as a way of challenging the cautious liberals of the Scottish establishment who caused them so much political grief when they were in power.

All these versions of scepticism about the governing system of civic Scotland — nationalist, radical left, and radical right — point not only to the potential for tension between it and the Parliament, but for the possibility of the reappearance, in new guises, of the old educational arguments against having any Parliament at all. As we will see more fully in Chapter 5, the defence of civic institutions will probably return to challenge the authority of the Parliament to intrude on professional concerns. Such arguments lost in 1997, but have not gone away. The conclusion of the present Chapter is that the inheritance from three decades of ardent campaigning for a Parliament is not so straightforward as many of these campaigners would believe. There are many good reasons to believe that the education system and the Parliament will thrive off each other, but there are also reasons to doubt that everything will be amicable. If the Parliament is really going to reform civic Scotland rather than just become another mechanism by which it continues to exercise the power it has enjoyed throughout the Union, then it will have to engage in some very controversial battles indeed.

Chapter 4

WHAT WILL THE PARLIAMENT DO?

Much of the debate about a Scottish Parliament's impact on education has been fairly abstract, couched in the very general terms that we examined in Chapter 3. That is understandable. In a reform that seems to reverse a central feature of Scotland's government for the last three centuries, paying attention to the minutiae of current policies may seem trivial. Nevertheless, it is in debate about policies that governments are tested. That is why, when asked in opinion surveys to rate the importance of 'issues', people have replied that a Scottish Parliament as such matters much less than particular social policies, not just on education but also on health, unemployment and so on. A Parliament is a means to social policy ends, and these have changed many times during the century or so in which a Parliament has been seriously debated.

We should not expect this to change now that the Parliament has been set up. The things which still dominate political debate will be particular policies, not abstract constitutional principles, or theoretical analysis of the relationship between the Parliament and civil society. Moreover, the influences on which particular issues will matter most will be long-term national and international trends rather more than the creativity of even the most inventive of Scotland's politicians. The Parliament does not start with a blank sheet, but in most areas of its remit takes over a set of policies and policy proposals that have been at most tenuously influenced by the prospect of national self-government. General trends have been important throughout the Union, as we summarised in Chapter 1 and illustrated in Chapter 2. Scottish education had to respond to industrialism, to mass democracy, to the welfare state and to its legacy of social rights just as did education systems in many other places. The Scottish response was distinctive — greater than average social mobility at the high point of the industrial revolution, greater than average preference for community and public provision in the era of the welfare state — but it was distinctive for being a unique combination of elements that were common throughout Europe and North America (including, it must repeatedly be emphasised, England). Responses to international currents will be the Parliament's main preoccupation too.

So this Chapter looks at the broad trends of policy which are likely to dominate the Parliament's deliberations, at least for the next decade or so.

Before that, though, we discuss the immediate legacy of the histories we have analysed in Chapters 2 and 3 — the legacies of educational developments in the unreformed Union (especially in the last three decades), and of educational aspirations that featured in the campaigning for a Parliament. This Chapter therefore concerns debates about the content of policy. The political and institutional implications of these debates are in the concluding Chapter 5.

The Immediate Legacy

Four features of recent educational, social and political change will shape the Parliament's impact on education. Each of these is, in some form, shared with many other societies, but each also has special Scottish aspects. Some of them were already being responded to by the self-government campaigning of the 1980s and 1990s, and so the themes overlap with some of those in Chapter 3.

The first is simply expansion, which has been happening at a more striking rate in the last three decades than at any time since the development of secondary education in the 1920s. It can be illustrated by five comparisons, relating to five stages of education from pre-school to university. The comparisons show that expansion continued as rapidly under the Conservative government after 1979 as it had in the previous decade and a half: the system was responding to deep social changes much more than to political initiatives:

- Until the 1970s, formal nursery education was to be found only sporadically. In 1971, just 8% of four-year-old children were in nursery education. By 1986, this had grown to 40%. By 1994, it was 55% (SED, 1977, Table 2; SOEID, 1995, Table 4; Swanson, 1975; Watt, 1990).
- In 1965, the proportion of the age group who did well enough in their Ordinary Grade examinations at roughly age 16 to achieve five or more passes was 22%. In 1981, it was 37%. In 1997, the proportion gaining the equivalent in Standard Grade was 56% (SED, 1991, Table 1; SOEID, 1998a, Table 5).
- In 1962, 15% of the age group stayed on beyond the minimum leaving age (which was then 15). In 1983, 52% of the age group stayed on beyond the minimum leaving age of 16. In 1996, 70% did so (Gray et al, 1983; SOED, 1994, Table 2; SOEID, 1998a, Table 2).
- In 1965, the proportion of school leavers who gained 3 or more Higher Grade passes at ages 17 or 18 — usually regarded as the threshold for entry to higher education — was 12%. In 1981, it was 20%. In 1996, it was 30% (SED, 1991, Table 1; SOEID, 1998a, Table 5).
- In 1961, the proportion of the age group which had entered full-time higher education by the age of 21 was 9%. In 1981, it was 18%. In 1997, it was 47% (Committee on Higher Education, 1963, p.26; Scottish Executive, 1999a, Table 12).

Many other examples could be given — for example, the expansion of community education since the 1970s, of further education colleges in the 1980s, of part-time post-graduate education since the early 1990s, and (more nebulously) of informal learning on the Internet — but the story would be much the same. The sheer scale of this expansion, in Scotland as elsewhere, has helped to give substance to the notion of 'the learning society'. This 'knowledge revolution' (another vogue phrase) has become one of the dominant cultural phenomena of the age.

The second legacy, in Scotland, is the implicit philosophy which has run through this — what we can call the comprehensive principle. As we saw in Chapter 2, the ending of selection for secondary school was put in place fairly swiftly and with relatively little acrimony in Scotland. The new system also remained popular, and so the principles on which it is based came to influence attitudes to policy in other areas of education. These principles are the particular current form in Scotland of the ideals of social justice that have been around since the founding of the welfare state, and which — as we also saw in Chapter 2 — gave coherence to Scottish opposition to the Conservative governments in the 1980s and 1990s. Generalised in this way beyond secondary schools, the comprehensive principles can be summarised as this:

- that people's access to education should, as a matter of legal right, be independent of their social circumstances so far as is practically possible;
- that the first point is consistent with continuing to hold effectiveness as a valid goal of the education system;
- and that, therefore, it is quite possible for an education system to promote both excellence and equal opportunity at the same time.

The reason why riders like 'so far as is possible' have to be added is that schools and other educational institutions are not in full control of the influences on students' abilities, and so cannot determine fully the reasons why a student is or is not in a position to take advantage of the opportunities that might be made available.

Principles of this type can be found in all developed education systems now. What makes Scotland distinctive (although not unique) is the relative importance they have had. Part of the claimed justification for expansion throughout the developed world — including Scotland — has been the pragmatic belief that education can serve the economy. This is the so-called 'human capital' view, in which investment in education is analogous to investment in machinery. The human capital argument is strong despite there being good reason to doubt that education really does have straightforward effects on the economy in the way that it claims (Ashton and Green, 1996). But in Scotland — as in Scandinavia — social justice has figured prominently as well. It may be that one of the reasons for this is some degree of popular appreciation that the human capital rationale is only partly valid. Parental opposition to national testing in primary schools seemed to be based on an idea that education should be about educating the whole child. A similar point can be made about the scepticism from the inspectorate as well as teachers about the crudest versions of performance indicators: the Scottish inspectors have been international leaders in developing indicators of 'school ethos', although it should be said that this in turn is often claimed to be justified on the grounds that it promotes better attainment.

Despite this support for the essential philosophy of the system that has been created over the last few decades, there is also now the consequences of the incipient dissatisfaction with it that we noted briefly towards the end of Chapter 3. This is the third important legacy. Beginning to appear from behind the façade of defensive unity that Thatcher provoked, it can be found in two forms. On the one hand there is the political right. As we noted in Chapter 3, freed now from the legacy of opposing the very existence of the Parliament, they are inclining to see it as a potential means of taking forward the reforms which the Conservative government of 1979 to 1997 tried and failed to impose. They start from a belief that Scottish education is decaying — far removed from the alleged glories of its golden age in, say, the nineteenth century, or in the days before selection for secondary school was ended. That decline dictates, it is claimed, radical change, but the problem is compounded by the system's reluctance to contemplate that: where educationalists admit that problems exist, it is alleged, they propose to deal with them by the same cautious gradualism that they have deployed throughout the period of the welfare state (or perhaps for even longer).

In England, that kind of view has always been influential on the debate about comprehensive education, aided by the rise of the Conservative party in the late 1970s. The so-called 'Black Papers' on education more-or-less formed the basis of the educational beliefs of Sir Keith Joseph and hence of Margaret Thatcher. The Scottish history of relatively successful and acceptable comprehensive schooling that we noted in Chapter 2 made the Black Paper critique somewhat irrelevant for the Scottish right, and so, by the late 1990s, they had to start their arguments afresh, on specifically Scottish grounds.

They could certainly quite readily find evidence that seems to back their case, because there is no doubt that the distinctiveness of the Scottish system has declined (and so that Scotland has come to share in common problems), although that mostly has more to do with other systems' catching up than with Scottish failure. For example, in the 1860s, the Argyll Commission found that the ratio of the number of university places to the size of the general population was about 1 in 1000 in Scotland, about 1 in 2600 in Prussia, and about 1 in 5800 in England. The Scottish ratio remained above the European average until after the Second World War, but it was then overtaken by the rapid expansion in many other systems from the 1960s onwards. Scotland now has rates of participation in higher education that are similar to those in such places as Denmark and Ireland, rather higher than in England, but much lower than in the USA.

There has also been a tendency by right-wing critics to compare the reality of the comprehensive system not with the previous overall reality of a selective system, but with a mixture of the most idealistic versions of the comprehensive ideal and the most successful products of the old system. Thus the failure of comprehensive schools to end social class differences in attainment is compared to a utopian aspiration that they could, and with the undoubted fact that many individual, academically able working class children (especially boys) managed to perform sufficiently well to make very effective use of selective schooling. As was mentioned in Chapter 2, for working class pupils who did manage to survive the many social hurdles to their taking Higher grade examinations at age 17, the system did manage to operate in a highly meritocratic manner.

Some of the evidence of decline that is cited is more controversial. There have been disputes over how to interpret the regular reports of the Assessment of Achievement Programme, which appear to show that attainment in the early secondary years has not been growing even though performance in primary has improved. There was disagreement between the inspectorate and *The Scotsman* newspaper about how to interpret the inspectorate's four-yearly report on standards and quality in schools issued in January 1999 (SOEID, 1996, 1999b): the newspaper emphasised the deficiencies that were identified; the inspectorate emphasised that systems were in place which they claimed would deal with these.

There have also been heated disagreements over the third international study of attainment in mathematics and science (SOEID, 1997), which showed Scottish children in the fourth and fifth years of primary school performing at no more than average levels (out of 21 countries, 13th in mathematics and 11th in science). The response from the inspectorate that the problems could be solved by gradual change has again been castigated by commentators on the political right, for example in *The Scotsman.*

What matters here are political perceptions, not academic disputes over how the international comparisons are to be interpreted. The validity of the comparisons have indeed been disputed: for example, the Scottish children in the samples were, on average, almost two years younger than the children in some of the countries that came near the top of the list, even though the same test was given in every country. When the performance of children who were all aged 9 was compared, Scotland appeared in a much better position: in mathematics it came 8th out of 21, and in science it was 5th. Scotland's ranking varied across particular aspects of mathematics (although not science): it was relatively poor in whole numbers (16th) but very good in geometry (2nd). It has also been pointed out that such international studies, because they rely on standardised tests, never measure some of the things that Scottish education has been trying to develop in recent years, such as oral skills, confidence at working in groups, and entrepeneurship. But this academic caution has received little attention, and the assumption of decline has become commonplace. In the same ICM survey as showed strong belief that the Scottish Parliament should control education (Chapter 3), as many as 27% believed that standards were lower than when they themselves were at school, 36% believed they were higher, and 26% believed they were the same (*The Scotsman Education*, 30 September 1998, p. 4). That suggests that the belief that the system is declining is quite widespread, although it is nevertheless still in a minority.

Much of this critique — but not the rejection of comprehensive education — is shared by influential people in the Labour party. They have argued that the comprehensive principle has to be rigorously renewed. For example, that seemed to appeal to Helen Liddell when she was briefly Scottish education minister in the year before the Parliament was elected. Further to the left than the Labour leadership, a less visible source of dissatisfaction is probably growing, resurrecting the educational radicalism which lay behind some of the comprehensive reforms in the first place — the ideals of people such as R. F. Mackenzie. This fits well with the current of leftist thinking which argued that Scottish civic elites are arrogant and complacent, and need a Scottish Parliament to shake them up. People of this persuasion can point to persisting social inequalities in access to education in support of their views.

The dissatisfaction with the education system has some sources in the current party-political debates in Scotland, among the Labour party, the SNP and the Conservatives. But it also relates to much deeper cultural changes, the fourth legacy which the Parliament inherits. This is a democratisation of culture and society — a reduction in deference, a more consumer-like attitude to public services, and a desire to participate in all sorts of decisions which used to be left to experts. Democratisation has amounted to profound changes in popular consciousness, 'in prevailing basic values concerning politics, work, religion, the family, and sexual behaviour' (Inglehart, 1990, p. 4). We do not have to accept the whole of Inglehart's thesis that 'post-materialism' is now the dominant cultural value in developed societies to agree that there has been an important change in political expectations. In the words of Anthony Giddens (who has been developing this theme for over a decade),

> In a society where tradition and custom are losing their hold, the only route to the establishing of authority is via democracy. The new individualism doesn't inevitably corrode authority, but demands it to be recast on an active or participatory basis. (Giddens, 1999, p. 66)

In consequence, people's political behaviour has become 'more demanding of autonomy and control' (Halpern, 1995, p. 344). The old Left failed to appreciate why the key elements of New Right ideology had such appeal from the 1970s onwards, even in societies such as Scotland where New Right parties were not very popular. By encouraging greater social mobility, and by making popular a language of individual rights, the welfare state itself had helped to individualise culture and ideology.

One reason why the New Right was able to promote this democratisation was simply economic. The economy internationally has required more flexibility, and that has influenced styles of governance in the public sector, the availability of qualifications for a wider range of skills, and the participation in the labour force of large numbers of women. But, as in specifically educational debates, this general dissatisfaction with an old style of welfare state has had sources on the left as well as on the right. The New Left of the 1960s in fact predated the growth of the New Right, offering a libertarian critique of the welfare state that, in some key respects, resembled the Right's. The New Left challenged the welfare state to fulfil the liberatory goals of its founders: state welfare may have been the necessary way by which collective oppression was alleviated, but the point of doing so was not to produce a rigidly collective emancipation. Freeing people from old constraints should mean freeing them to be as different as they wanted to be. Thus the New Right was not the only political movement to benefit, and the phenomenon of 'new social movements' has been attributed to the same sources. Collective interests other than the Labour movement could use the ideology of the welfare state to assert their rights too.

Scotland shares fully in the persisting support for the welfare state, as we have seen. Scots are also at least as favourable to more popular participation and influence as people elsewhere. There has been a decline in deference as old social institutions have lost their authority, notably the churches but also, probably, the leading professionals who have been accustomed to running public services such as education with such undisturbed authority. People in Scotland, as we mentioned in Chapter 2, have been quite keen to take advantage of greater freedom of choice in public services, even when that was put in place by an unpopular Conservative government (and even when the exercise of choice might undermine the principles of social justice which remained the highly popular basis of the welfare state). So this legacy for the Parliament is as difficult as that faced by political institutions in other broadly social democratic countries throughout Europe — how to maintain the principles of social justice while also welcoming and promoting greater individualism.

The Parliament's Response

The Parliament will not be powerless in the face of these trends, but it would be as ill-advised to disregard them as the Conservatives were when they tried to ignore Scottish predilections for social justice. Most obviously, the expansion that has been inherited will provoke demand for more: large amounts of research have now shown that the more education people have, the more they want. This also works between generations: the more education parents have had, the higher the educational aspirations of their children. Thus the Parliament will simply preside over the next phase in the growth of higher education, probably using the further education colleges to a much greater extent than hitherto. It will continue to expand pre-school education because working parents want it as a high-quality form of child-care (and because there is some evidence that it raises educational attainment later on). They will further expand the rate of staying on beyond age 16 in schools, because the Higher Still reform of the examination system for that age group is now beginning to be put in place. And they will encourage lifelong learning — and the community education which supports it — because the well-educated adults who benefited from the first phase of expansion in the 1960s and 1970s are now returning for more.

All of this will be affordable, in broad terms, not because Scotland is unique but because it is not. The block grant that provides almost the whole of the Scottish Parliament's funding will allow Scottish educational spending to grow roughly at the same rate as analogous budgets in England, because higher education, pre-school education, staying-on and lifelong learning are all expanding there too. Rather ironically, the Parliament will almost certainly be successful in simply bringing more education to Scotland because more education is happening in the rest of the Union. Unionists might see that as a vindication of home rule as against independence, although it could equally show that any government does no more than guide the most important social changes.

The Parliament may also respond to the popular desire for diversity and influence by reforming the character of public education. Higher education is already more diverse than school education, and is likely to become even more so, as some universities specialise in research, others make a virtue of their vocational links, and the further education colleges warm to the idea that they are the best way of promoting open access. That is diversity of content. Pre-school education has expanded so rapidly that it has had to be diverse in its governance, a mixture of public, voluntary and private. That diversity has caused diversity of provision, the public nurseries being more tightly tied to the curricular specifications of the inspectorate, and the private ones placing more emphasis on the combination of education and day-long child-care. Even schools — and even in Labour local authorities — have become more autonomous and are probably becoming slowly more diverse. Devolved school management has, on the whole, been popular. And some Labour authorities are pioneering diversity: for example, Glasgow is proposing to encourage schools to specialise in particular areas of the curriculum. There is some political pressure in the Parliament — and even in the Scottish Executive — to reform governance further by slowly giving more autonomy to local government, in line with the recommendations of the report of an official committee (Scottish Office, 1999), and in keeping with a frequent claim of the self-government movement that devolution had to go beyond the mere setting up of a Parliament in Edinburgh.

Examples such as these indicate how politicians can influence the ways in which an ineluctable trend has its impact. Expansion will almost certainly happen, but how it happens will depend on how politicians choose to relate that trend to the equally strong trend towards greater diversity. Scottish politicians will claim, of course, that one area in which they are making a unique contribution is in their renewal of the popular ideals of social justice (the second legacy). That will appear plausible because they are still able to contrast this with the hostility towards these principles that came from the Conservative government, and also with an apparently weaker commitment by Labour in England (although that latter contrast has to be muted when it comes from the Scottish Labour leadership). In some respects, these claims will be specious, if the argument advanced earlier in this chapter and in Chapter 2 is correct: social justice is a popular Scottish principle to which politicians have to respond if they want to be popular too, and so the credit really belongs to the society at large, not to politicians' leadership.

Nevertheless, politicians can help to focus principles on new institutions. When he spoke at the opening of the Parliament on 1 July 1999, the Scottish Labour leader Donald Dewar pointed to the inscriptions on the Parliament's mace — 'wisdom, justice, compassion, integrity' — the first two of which link education and rights. That attaches the idea of social justice to the Parliament itself (in a speech which, in its evocation of an age-old Scottish moralism, belonged as firmly to that version of Scottish nationalism which finds ancient sources of Scottish democracy as it did to international social democracy). Political thinking is also now beginning to find scope for renewing the principle of justice in ways that are quite innovative. The New Community Schools are an example. Based on the idea of full-service schools in the USA, they aim to bring a variety of social and health services onto the same site as the school in order to deal directly and efficiently with the various social barriers to learning. The purpose is to leave teachers free to be educators, rather than surrogate social workers. A second example of innovative thinking — this time from within the traditional policy community — is an official report on community education by a committee chaired by the head of the inspectorate, explicitly linking it to social justice and even to promoting social change. Another instance is the role of further education colleges in expanding access to higher education. Because they already have a large share of higher education students, they are likely to be seen as a way of radically widening access without upsetting the research aspirations of large institutions such as the ancient universities.

These examples also indicate how Scottish politicians tend to respond to the allegation that the education system has weaknesses (that is, to the third immediate legacy). They observe that schools serving middle class pupils, and the older universities, seem to have high academic attainment, and then argue that interfering structurally with them is pointless. Although they instigate curricular reforms that are intended to stimulate the most academically able students — for example, a new Advanced Higher or a new high attainment level associated with the 5–14 curriculum — they trust the middle-class schools to put them in place effectively. All that is required is regular inspection to ensure that these places can develop their own responses to temporary weaknesses; anything more intrusive, it is believed, would risk destroying august institutions. According to this dominant view, government action is needed, however, to improve the quality of provision for people living in poverty — the 'socially excluded' in the current jargon. In short, Scottish politicians are inclined to re-interpret problems with standards as problems with socially unequal access to good-quality provision; this is an astute move politically, because it links the debate about standards with the debate about justice. It is the approach of New Community Schools, and of the main concerns of the proposed reforms in community education. It is also the focus of the most important initiative in education for ages 4–7, the 'early intervention' which encourages attention to developing literacy among children who live in poverty and who have only limited access to opportunities for learning outside the school. By linking action on standards to questions of justice, Scottish educationalists thus also appeal to most of the political left, and provide a means of apparently furthering justice while not appearing to be soft on quality.

Responding to a more individualistic culture — the fourth legacy — is partly a matter of making provision more diverse. It also is connected to the debate about citizenship, which we summarised in Chapter 3. Because that debate has a history on both the left and the right, as well as being part of most current thinking about constitutional reform, developing citizenship is a way of responding to a social trend while also keeping most political factions satisfied. Indeed, the syllabuses of Modern Studies in secondary school and college can be claimed to be already in a good position to become the core of a wider programme of education for citizenship. The Scottish Parliament's education service headed by a former teacher of Modern Studies, and the Modern Studies Association (with other organisations) ran a mock election in schools to coincide with the first elections to the Parliament in spring 1999. A Youth Congress has been set up, partly sponsored by the Parliament's education service. The official report on community education envisaged that service as having a role in educating adults in how to be effective citizens. All of these separate activities have in common their use of electronic media: web sites, email, conventional broadcasting. They have all also tried to learn from analogous experience in other countries, acknowledging that a popular desire to participate is indeed a very common phenomenon. So the Parliament's information services have drawn on experience in many democracies, the Youth Congress can draw on long-running similar bodies in several Canadian provinces, and the debate about citizenship education has paid attention to similar discussions elsewhere, including in England.

Thus Scottish politicians have shown themselves to be aware of the unavoidable trends that we looked at earlier in this Chapter, and in some respects they are responding quite imaginatively. They have drawn on the ideas of the existing policy community for action to deal with the more technical aspects of changing formal education. New Community Schools are a good instance. For less conventional topics, such as citizenship education, they have been more innovative themselves. No doubt a lot of this will continue to happen quite harmoniously. Politicians and a slowly widening leadership class will coexist and will, in due course, settle down into a reasonably constructive relationship. This prospect is assessed in Chapter 5. But it cannot be the whole story, because how to respond to the social and educational trends we have been looking at is not as uncontroversial as the illustrations we have looked at so far might suggest. To finish this Chapter, therefore, we turn to the tensions which will arise in connection with each of the four legacies that we identified.

The Difficulties of Responding

Underlying the difficulties is the tension between technocratic management and social justice which we looked at in Chapter 2, and which is the dominant theme of the final Chapter. One example is in the various justifications which have been offered for expansion, and the relationship of these to the ideal of social justice. We have already noted that both expanding human capital and achieving social justice were influential in Scotland throughout the twentieth century. Writing about British education policy in that period more generally, Raymond Williams (1961) adds a third, which he calls 'old humanism', the views of people whose philosophical inclination was that education should be universal, but who did not share the full social radicalism of the political left. Scottish examples are easy to find, precisely because influential segments of educational opinion have been inclined to believe the Scottish myth of social mobility through education. Because such people still dominated the Scottish Education Department, the inspectorate and the local authorities when the 1960s Labour Government was encouraging the growth of comprehensive schools, they provided a reassurance that this radical change was actually just a renewal of an old Scottish principle after all.

That class is still there, but its power has been eroded by a mixture of comprehensive education itself, the democratisation of culture we mentioned earlier, and Thatcherite intolerance of civil service caution. So its beliefs can no longer form a diplomatic bridge between the radicals and the advocates of human capital, and therefore the potential conflict between these is starker. The tension is a version of two ways of justifying equal opportunities — the mere absence of constraints, or the active public pursuit of social justice for whole deprived communities. Although the two versions of equal opportunities can get along well enough for a while, ultimately they clash, especially when there are finite resources.

To make this more explicit, take higher education as an illustration. The most effective and efficient route to widening access in Scotland would be to expand higher education provision in further education colleges. But if that route is preferred by the parliament, then it is bound to mean a net shift of resources away from higher education institutions. Not only will that be unpopular with these institutions themselves. More importantly in a political sense, it will squeeze some students out of them and into further education colleges. So, although whole communities could benefit considerably by the strengthening of higher education provision in the further education colleges in their midst, some individuals might feel their opportunities had been restricted, and the older universities, in particular, would be likely to complain that their venerableness was being threatened. The resulting controversies would resemble the disputes of the 1960s over comprehensive secondary schools, and the fact that these debates were resolved in favour of a unified system that became and remained popular suggests the direction that the debate about higher education is likely to take.

Expansion and social justice, then, are not so easily reconciled as social democratic reformism might claim. It is much easier to manage the burgeoning meritocracy of a system of individual rights than radical claims that deprived communities are deprived precisely because rich communities are advantaged. Rhetoric of class conflict may not be as fashionable now as it was when comprehensive schooling was being introduced, but it is still vociferous in Scotland.

There are also, very obviously, controversies over the very meaning of standards, quite apart from the conflict over how to raise them (to which we return in Chapter 5). The controversy is analogous to the two versions of equal opportunities. Is educational quality about meritocracy, and therefore highly competitive? Or is it about success for all? The first view tends to define standards as being about the difficulty of the tests which everyone undergoes. So equal standards for all mean that everyone may proceed through the same graded levels (even though most will fail or opt out on the way). That is the meaning which is invoked when people claim that the difficulty of public examinations is falling (or is being maintained or is rising). The second meaning of standards — tending to be favoured by those who believe that success for everyone is possible — is the proportion of the population who reach a certain level, so that the highest standards are attained only when everyone reaches the same high level. That meaning is invoked when people cite as evidence of rising standards the proportion of young people going to higher education, or when they cite as evidence of falling standards the declining proportion of school students who pass public examinations in modern languages.

It is easy to find rhetoric which claims that an education system can have high standards in both senses, and to some extent it can, especially at times of overall expansion, but ultimately such claims merely evade the difficulties. One way of expressing the problem is to point to the sieving which an educational system has to do. So long as different occupations demand different levels of education, there is no escaping that the education system sorts people for jobs (even if, nowadays, the sorting has to take place again several times during a person's career). The highest levels of attainment will simply not be needed for many jobs, and so the education required to reach these levels is not likely ever to be afforded for everyone. In that sense, societies with complex systems of occupation accept that there is a limit on the improvement in standards defined in the second sense — as the proportion reaching each level.

These confused interpretations lurk somewhere in the background of the arguments about standards, and ensure that there is now no consensual way of resolving the debate. In the period from the 1930s until about the 1970s, there could be agreement between educational reformers and educational conservatives that, regardless of the organisation of schooling, standards should remain high in the first sense: the difficulty of examinations should not change. But when reformers ask for sufficient investment to overcome all the reasons why some people fail at some point — aiming to raise standards in the second sense — then they and conservatives simply cannot agree. Moreover, if the contrast was as stark as that, it might be easy to take sides, but it is not. Educational investment really does increase the proportion of young people who can reach a given level (standards in the second sense). Thus, even if there has been some fall in the difficulty of the Higher Grade examinations (standards in the first sense), even educational conservatives do not claim that that entirely explains, say, the rise in the proportion of the age group gaining Higher English by the time they left school from 13% in 1962 to 36% in 1994. The controversy is not over whether some people can be educated beyond where they are at the moment. It is over the limits to that process, both for each individual and for the population as a whole.

That is why arguments over standards never go away, and why they relate to the other controversies which the Parliament inherits. The dominant proponents of comprehensive education have wanted high standards all round — standards in the second sense — as a means to social justice, and they have believed that these can be achieved by appropriate levels of investment in appropriate structures of education. Reforming conservatives believed that educational selection of some sort is forced on an education system by occupational selection, and so that the fairest way of maintaining standards is by paying most attention to the quality of the tests. Such views have not been common in Scotland recently, although they were very widespread when they formed the basis of the selective system in the 1940s. Recently, even segments of the English left have come to accept them — for example, the journalists Will Hutton (1995) and Peter Preston (1999). Thus, no matter what the Parliament does about standards, it is bound to be severely criticised by one side or the other. If it seeks compromise, it will be lambasted by both: investment will be alleged not to be enough to raise attainment by everyone, but continuing slow increases in attainment despite a lack of investment will suggest that examinations are getting easier.

There are controversies, too, about governance. The fact that there has been a democratisation of culture tells us nothing about the most valid response. Policies which imagine the citizen to be a consumer are not necessarily in conflict with policies which aim to empower communities, but they certainly can be. Merely asserting that the Scottish Parliament can improve the accountability of public services in Scotland glosses over the different meanings of 'accountability'. Is it the accountability of the market? Is it technocratic accountability to managers further up a hierarchy, or to external auditors and inspectors? Or is it classic political accountability to whole communities, whether locally or nationally? The implications of the choice can be radically different, even incompatible. Consider higher education again. Customer accountability seems to imply a system of real fees, set by individual universities in response to demand. The USA provides plenty of successful examples, but this would not be likely to be politically acceptable in Scotland. Managerial accountability means tight monitoring in the form which has become familiar in all branches of education in the last two decades. That may be quite acceptable to the top end of the hierarchical Scottish policy community. In that case, it would be difficult to see how it could be acceptable to a Parliament dedicated to shaking up the leading components of that policy community, although it might be quite welcome to politicians who would be inclined to ally themselves with these components against the politically weaker parts of the education system — for example, in this instance, lecturers in further education colleges. Community accountability would mean some direct community involve-ment in university government, and if it was not to be a merely local version of the hierarchical leadership class it would have to be by direct, popular election. That is unlikely to be acceptable to the existing universities, or to the further education colleges who welcomed being removed from the control of elected local authorities in 1993. The degree of local democratic legitimacy which this would confer might not be acceptable to the politicians in the Scottish Parliament either, because it might challenge their power.

For similar kinds of reasons, there are controversies over the meaning of citizenship. On the one hand is the tradition of 'civic republicanism' and 'civic virtue' — the idea that education should make people into responsible citizens. In the words of the report of Crick's committee in England,

> The purpose of citizenship education in schools and colleges is to make secure and to increase the knowledge, skills and values relevant to the nature and practices of participative democracy; also to enhance the awareness of rights and duties, and the sense of responsibilities needed for the development of pupils into active citizens; and in doing so to establish the value to individuals, schools and society of involvement in the local and wider community. (QCA, 1998, p. 40)

The emphasis here is on responsibilities and duties, and there is no explicit encouragement to question the norms and values of society, or to challenge the government. This view fits well with the communitarianism that has been influential on the Blair government. It also could be quite consistent with the more authoritarian aspects of traditional Scottish culture, a lot of which probably survives despite all the democratisation. Nearly all the Enlightenment thinkers, after all, linked their enthusiasm for ancient Greek republicanism with a staunch presbyterian Protestantism.

On the other hand, there is also a more dissenting interpretation, in which students are encouraged to challenge authority. English debate has also produced this interpretation, as distinct from Crick's:

> for citizenship to flourish, there must always be a critical tension between the ideals and values in terms of which the rights and duties of citizenship are defined and the dominant institutions which make provision for their practical enactment and realisation. (Carr, 1991, p. 380)

Such views can in fact be found in the Scottish Curriculum Council's review of Scottish culture that was mentioned in Chapter 3. Young people, it said, should be encouraged 'to make their own judgements on what is more or less valuable in the cultural life of the country' (SCCC, 1999, p. 16). Insofar as citizens are autonomous, citizenship is always defined against the state as well as in support of it. The same conclusions can be reached by noting that a society made up of diverse cultures — religious, ethnic, regional, and so on — ought not to expect all its citizens to interpret social responsibility in the same way.

Conclusion: The Inevitability of Conflict

None of these several tensions over the interpretation of social and educational trends will be easily avoided. Despite the claims that the Scottish Parliament would inaugurate a less confrontational style of debate, each side in each of the arguments is very ready to claim that theirs is the only way forward. This self-assurance regularly appears in, for example, editorials in *The Scotsman*. Criticising the caution of the education white paper which was issued by the Scottish Office in January 1999, the newspaper said:

> Comprehensivisation ... has certainly not met the objectives set for it. It is time to acknowledge that it never will. Until politicians discover the courage to say this ... Scottish schools will be doomed to perpetual mediocrity. (*The Scotsman*, 28 January 1999, p. 16)

We find a similar confidence in an opposing direction from the Educational Institute of Scotland, in the manifesto it published just before the elections to the Scottish Parliament in 1999:

> Scotland has an education system to be proud of. (EIS, 1999, p. 20)

And each of these castigates the other with a vehemence that would have not been out of place in the great nineteenth-century debates within Scottish presbyterianism:

> Debate about education in Scotland has been stifled for too long by the dominant role of an educational establishment which adamantly refuses to recognise truth. (*The Scotsman*, 28 January 1999, p. 16)
>
> Those who would seek to belittle the system of education in Scotland … have distorted schools inspectorate findings, built on ambiguous international comparisons, and tried to adapt media comparisons of problems in schools in England as if they applied to Scotland. (EIS, 1999, p. 20)

These are not artificial divisions, however much the rhetoric may seem rather overblown. They reflect the importance of the themes of tradition, expansion, standards and equal opportunities which we have been analysing, and they reflect also the impossibility of reaching a universally agreed view about how best to react to them. The belief that there could be such a view was the most insidious legacy of the long years spent campaigning for a Scottish Parliament — the illusory unity of the anti-Tory front. The implications of the ending of the apparent consensus are the most interesting political question now facing Scotland. That is the subject of the final Chapter.

CHAPTER 5

THE PARLIAMENT AND CIVIC SCOTLAND

Two futures are possible. In one direction, the Parliament and the civic institutions of Scotland get on very well together. In the other, they compete for leadership. Education will be at the crux of the debates, either way. Either it will be the supreme test of the Parliament's capacity to reform the sclerotic structures inherited from the unreformed Union, or it will lead the civic institutions' heroic resistance to political intrusion. It will have these roles because it is at the heart of Scottish identity, in the ways that were traced in Chapters 1 and 2. Of course, this opposition is too sharp. These are not alternatives: both will be true, in part, because neither Parliament nor civic Scotland is homogeneous, and the two prospects for Scottish government will run into each other at every turn.

This Chapter looks at these two versions of the future in more detail, tracing the influences which are pushing in each direction. As throughout the book, the theme of the Chapter is how policy is made, not directly its content, but the different styles of policy process which could develop are illustrated for each of the four elements of the legacy that was identified in Chapter 4 — expansion, the comprehensive principle, questioning of standards, and greater individualism. It does not try to assess which of the two versions will dominate, far less to say which would be best (whether educationally or democratically). The main point is that the relationship between the Parliament and education will not be the harmonious unity which a century or so of home rule campaigners have imagined.

Working with the Civic Institutions

The strongest influence towards harmony will be the pressure towards so-called new politics that accompanied the development of the Constitutional Convention's scheme. This is the preference of the politicians, as embodied in the report of the Consultative Steering Group and the Parliament's resulting standing orders and working practices. It also seems to be to the liking of the leaders of Scottish education. After the confrontations of the Thatcher and Major years, there was a widespread sense that education had had enough of politics altogether. The Parliament, oddly, seemed a political route to getting rid of political acrimony.

The Parliament and the Scottish Government will no doubt, therefore, try to renew Scottish pluralism by forging new ways of liaising with civic

Scotland. For the Government, this will involve the rather delicate task of engaging in wider discussions beyond the civil service and the inspectorate while also not annoying them: thus — for all the aspirations to very wide consultation — new pluralism is bound to find a new home for the old leadership class, if only because they will continue to be the main way in which Ministers will find out what civil society is thinking. The standing orders of the Parliament require that draft legislation should be accompanied by an explanation of its rationale and should seek responses; the first Education Bill, published in July 1999, was an early example. If Ministers really do listen to what is said, then a wider range of civic bodies is likely to have an influence on legislation. The reasons to believe that Ministers of the first coalition government will do this is a mixture of the attachment of individual Labour Ministers to the ideals of consultation, the quite good record of consultation in local government by the Liberal Democrats, and the likely tendency of opposition politicians and the Scottish printed press to portray every disagreement between civic Scotland and the Government as a betrayal of the ideals on which the Parliament was supposed to be founded.

For the Parliament, as opposed to the Government, the pressures towards consultation will be most strongly associated with the system of specialist committees, proportionately balanced as between parties in just the same way as the Parliament as a whole. There are two with an educational remit — one dealing mainly with school education, the other with higher education, further education and lifelong learning. Their most important job will be to seek evidence, and evidence will come most easily from civic bodies with a long record of lobbying. Educational interests abound in this respect, precisely because they were used to being consulted in the unreformed Union, and became indignant during the long period when the Conservative government paid them little regard. Again, the opposition parties will find the committees a convenient forum for probing any disagreements between the Government and civic Scotland, and the news media will find that inquisitorial committees will probably make for rather more exciting stories than most plenary debates (just as in the US Senate, or even with the select committees at Westminster). It may be that the fairly high proportion of women in the Parliament will also encourage constructive dialogue with civic bodies outside, in the manner envisaged by feminist campaigners: 37% of the members are female (including one half of the Labour group and 43% of the SNP's).

Some of this pressure for consensus between the Parliament and civic Scotland may be expressed in a Civic Forum. The idea for this had its origins in a Civic Assembly set up in 1995 by the Scottish Trades Union Congress, and involving representatives of trades unions, churches, pressure groups, and other non-governmental organisations. Some of its advocates imagined that it might act as a kind of civil-society conscience for the new Parliament, and they proposed constitutional mechanisms by which ideas from the Assembly could be fed into parliamentary debates. None of these proposals appear in the legislation setting up the parliament, but a Civic Forum was proposed by the Consultative Steering Group (borrowing the name from the similar ideas in the 1998 peace agreement in Northern Ireland). A Forum has in fact been set up and will receive some state aid, and there will probably be some requirement that the Parliament and the Government should consult it. For all the rhetoric of participation which surrounds such a body, however, it will mainly embody consultation with already active civic groups. Partly because of a fear that the Forum will thus institutionalise consultation, there have also been discussions about maintaining an unofficial Civic Assembly that would have a broader membership. Searches like this for the locus of true popular accountability illustrate that the Parliament was already — within a year of the first elections — being felt to be unable fully to realise the democratic ideals on which it was founded.

Education is likely to be debated quite often by the Forum and its penumbra of consultative bodies. There may also be some kind of Education Convention, modelled on a successful example in Ireland in 1993. It took evidence nationally and regionally, and its final report became the basis of a series of reforms that were broadly agreed among the political parties. The SNP has already supported the idea, and away from the heat of the election campaign (when the chief Labour spokesperson — apparently ignorant of the Irish precedent — dismissed it as a talking shop), it is the kind of civic idea which the Parliamentary committees may like.

There will be partnership with educationalists also in the process of implementing whatever legislation does eventually come from the Parliament. That is inevitable, as was argued in Chapter 3: policies have to gain the cooperation of professionals if they are to have any impact at all. It is also to the liking of politicians inclined to consult. After all, most people will experience the Parliament's impact not directly but through the ways in which various public-sector professionals interpret its proposals.

A second source of pressure forcing the Parliament to work with civic institutions is the inexperience of the new politicians, and the power of the inspectorate. The inexperience was evident in several very noticeable incidents during the first few months of the Parliament's operation — for example, the rather inexpertly handled debate on how much financial support the members should have for their office and research costs, the apparent lack of historical imagination in connection with the proposed new building for the Parliament, and the failure to get to grips promptly with the operations of professional lobying firms. It was also evident in some of the debates, for example on tuition fees, in which the principles of consultation, compromise and the careful preparation of policy seemed to have been abandoned in favour of partisan rhetoric. Thus the opposition parties wanted fees to be abolished immediately. The supporters of the Government claimed that the committee of inquiry into student finance was an example of rational policy-making. The opposition alleged it was a political fix that allowed Labour and the Liberal Democrats to reach quick agreement on a coalition, and said that rational policy making — especially involving consultation — required much more than the six months which the committee was allowed. There was also a sense from several contributions to the debates that members were not well-briefed, plucking statistics out of the air to suit their case: this sense was as strong after the committee had reported as it was when it was being set up. There was no awareness, for example, of the impact which social democratic expansion had had on widening access to education — a process that had continued under the Conservative government — and little sense of the difficulties of setting priorities in a limited budget.

Inexperience will pass, and research support will help the MSPs to understand policy and precedent better. But in seeking appropriate advice on education to overcome their inexperience, the MSPs are likely to turn to that most traditional of sources, the inspectorate, who are likely to continue to play their role of advising debates and therefore of shaping policy. If they do become the main source of advice to the Parliament's committees as well as to Ministers, then the scope for real debate between the executive and the legislators will be substantially reduced. The ideas available to most MSPs would come largely from within the education system. This would tend to tie policy closely to the preferences of the already influential segments of the existing policy community.

The policy process that may emerge around the Parliament is only one reason to expect it to work closely with civic institutions. In the background, there will be the further reason that Scottish professionals are popularly trusted, for the many reasons that we looked at in Chapter 2. They know how to get things done. That was a reason why the welfare state came to be dominated by professional groups in the first place, and it remains a reason now that the standing of politicians has fallen. Politicians may be partisan, self-interested and possibly nefarious, but teachers are trusted to place the interests of their students first. A profession which has extensive engagement with civil society is likely to command much greater credibility on educational topics than elected politicians, in Scotland as in many other places.

This is also likely to become the way in which original scepticism about a Parliament is re-expressed — not as persisting opposition to its existence, but as questioning of its right to do things against the wishes of the system. That position will be, as it were, the civic defence against the radical possibilities which a Parliament might bring — against precisely the kind of challenge which, as we saw in Chapter 3, intellectuals such as Tom Nairn pose from one direction and the new Scottish right from the other. Eliding the difference between teachers and the policy community, the widespread trust in teachers will be cited as a way of asserting the policy community's political legitimacy. It will be alleged that the policy community could be trusted much more than upstart and inexperienced politicians. It will be claimed that politicising educational debate would be a bad thing in itself — that the virtue of Scotland's peculiar position in the Union was precisely that it had control of the things that really mattered (the details of curriculum and so on) without having to bother about the messy business of politics. And it will be suggested that the professions, because linked to professionals in other countries, are much less parochial than politicians are bound to be, especially politicians coloured by some degree of nationalism. The likely popularity of such arguments will force the Parliament to respect the views of the leading segments of education, the policy community that governed education in the unreformed Union.

A final source of caution in the Parliament's activities will be an abiding sense that anything which questions the basic value of Scottish education is somehow anti-Scottish. That is one reason why the last education Minister in the old Scottish Office, Helen Liddell, attracted so much opprobrium by her plans for relatively mild reform. She was accused by the EIS and the SNP of being almost as hostile to the Scottish education system as the Tories had allegedly been, and as being as antagonistic towards teachers as Labour Ministers in England. An institution so close to the heart of Scottish identity is bound to be more thoroughly cherished than the new Parliament, at least for some time. To gain affection, the Parliament will be inclined to show that it is not hostile to the respected old symbols. Thus the draft Education Bill published in July 1999 makes frequent flattering references to the essentially self-renewing character of Scottish education. There is nothing wrong with the system's values: 'the commitment to excellence is fundamental to Scotland's educational traditions' (Scottish Executive, 1999, p. 9). The belief that the system is and ought to be autonomous is celebrated: 'Scottish ministers have no wish to reduce flexibility at local level' (p. 33) and '[local] authorities and schools' will be subjected to 'minimal central direction or intervention' (p. 15). And the leadership class cannot be bettered as the agency for overseeing any central involvement there might be: 'ministers consider that [the inspectorate] is uniquely placed to provide the overview of the quality of provision, and its management, based on their existing experience and expertise' (p. 35).

If anything, the opposition parties endorse this even more thoroughly. During the election campaign, the SNP went out of its way to be friendly to teachers and local authorities:

> The SNP in government will stop the endless political meddling in education. … The SNP will allow schools, in consultation with local authorities and school boards, to set educational targets that reflect their particular circumstances. (SNP, 1999, p. 21)

In its desire to appear respectable — and to gather votes from large professional groups — the party has eschewed the radicalism of writers such as Tom Nairn. Nairn, in turn, has frequently castigated the party's caution. He wrote in 1970 that:

> The belief that a bourgeois parliament and an army will cure the disease [of Scotland's cultural deformation] is the apex of lumpen-provinciality, the most extreme form of parochialism. (Nairn, 1970, p. 51)

And then he repeated the comment in 1991, explaining the apparently endless cycles of SNP rise and fall:

> the SNP is 'bourgeois' all right: ... the alter ego of the institutional bourgeoisie. ... An institutional elite deprived of an elite's most important characteristic — command power — is far more comfortable with a rhetorical than with a practical nationalism. (reprinted in Nairn, 1997, p. 189)

The failure of the SNP to win the 1999 election is — according to this view — therefore more than just a partisan disappointment; by keeping it attached to institutional caution, it postpones the party's growing up.

Even the Conservatives have been trying to reinvent themselves as the champions of the education system. The report of a committee on the future of the Scottish party, chaired by Malcolm Rifkind, admitted past errors of political interference:

> we recognise that many in the teaching profession feel disillusioned because of the authoritarian manner in which some Government legislation has been enacted. (Scottish Conservative and Unionist Party, 1998, p. 17)

and

> control can rest with those who are full-time educationalists as well as elected politicians. (p. 18)

Hostility to politics is exactly what Nairn and Humes said was wrong with the Union, as we saw in Chapter 3, and so none of the large parties — even the SNP — seems ready to offer the radical break with the old style of Union which such critics claimed was needed.

We can guess at the ways in which a Parliament dominated by this style of thinking will deal with each of the four dominant current trends that we identified in Chapter 4. On expansion, the inclination will be to let the system take the lead, and to produce a compromise whenever an inescapably difficult issue arises. The question of tuition fees is a striking instance which we have already looked at: the outcome is the same kind of managed agreement which the old Scottish Office was rather good at until the 1970s, and was occasionally still achieving well into the Thatcher period (such as over vocational education in 1984 and over national testing in 1991–2). Lifelong learning is another. The human capital arguments will dominate — that more adult learning is good for the economy — and so the potential tensions between that and more radical visions will be evaded.

On social justice, the approach will be dominated by the inherited sense that gradual reform works better than anything more radical. Gradualist social democracy gave us the fairly successful system of comprehensive schools, it will be argued, and so can equally deal with the necessary reform of these and with analogous issues concerning access to higher education. Further education colleges will be allowed to expand their provision of higher education still further, thus drawing in more working-class students, but in an ever-expanding budget this will not provoke protest from universities, because no-one will lose out in absolute terms; indeed, the universities will continue to attract onto full degree courses many of the students who complete diplomas in a college. The expansion, in other words, will be justified in terms of the competitive version of equal opportunities, as offering chances for individuals. Little will be said about the effects on communities: there will be little championing of the network of local colleges as offering a radically new way of defining what higher education means.

On standards, the response will be the gradualism inherent in the draft Education Bill. It comments on the controversies which were aroused in January 1999 in response to the inspectorate's report on standards and quality (as was discussed in Chapter 4):

> [The] report ... confirmed that as a whole the system performs well. It is a matter of regret that undue attention was given in public reporting to areas of weakness that were identified. [The] report gives [ministers] every confidence that any areas of weakness can be addressed and overcome and action is already in hand to do so. (Scottish Executive, 1999b, p. 11)

There will be targets to measure progress, but these will be set by educational managers and inspectors. Therefore they will relate to those things which the dominant people in the system value most — examination passes, rates of progression to further stages of education, coverage of the curriculum. These criteria will often be defended, no doubt, on the grounds that they are what parents and employers want too; but, by and large, they will not move towards any redefinition of the purposes of education, say in the direction that would be implied by the tradition of radical educational thought stemming from A. S. Neill.

The response to democratisation, finally, will be the educational instances of the changes to the policy process which were outlined earlier in this Chapter. Participation will still largely be by invitation and by pressure group. There will be no direct election to quangos or even direct political participation in them (say by MSPs or elected local councillors): the most that might happen is some formal process by which the Parliament has to ratify appointments, and occasionally hears evidence from candidates (as in the US Congress). Consultation will be dominated by pressure groups because that is how the whole structure of Civic Assembly, Civic Forum and Parliamentary committees is likely to work. The educational pressure groups will claim legitimacy because they will be able to show that their concerns are truly popular — that people do trust the professionals who staff and run the system. Their dominant ideology of gradual reformism will allow the Parliament to preside over further gradual expansion, and so the system's legitimacy will be further strengthened.

In other words, if the politicians — both governing and in opposition — take the cautious route they have favoured up till now, and which the rhetoric of consensus strongly encourages, then the Parliament will indeed renew Scottish social democratic pluralism, and will open up the policy circles beyond the narrow social groups which dominated them in the 1950s. There will be more women, more Catholics, more primary teachers, more people from ethnic minorities. Therefore there will be less of the Kirriemuir career. But that widening has been happening for some time, and the essential structure of Scottish civic life will remain undisturbed. It will still be civic, still strongly institutional, and still managed from the centre to keep popular politics out.

Tensions with the Civic Institutions

Some of these pressures inclining the Parliament to work with the civic institutions may start off as tensions. It may be that the true extent of the trust which is placed in Scottish education will only be tested when the Parliament tries to challenge the most influential segments of the policy community. It may be that the governing politicians will have to fail in some attempted reform before they conclude that working with the system will be more effective. That is, after all, what the Conservatives learnt in the early 1990s, after Michael Forsyth largely failed to shake up the local authority and Scottish Office roles in governing education. For example, it is conceivable that an MSP or the lifelong learning committee might table legislation to shift resources from the universities to the further education colleges. If the universities then successfully mobilised opinion against that — on the plausible grounds, say, that it would threaten their capacity to attract UK research funds or to attract students from outside Scotland — then such a radical political initiative would not be likely to be repeated.

So the really interesting tensions between the Parliament and the education system — the ones with the most fundamental implications in the long term — would be those which were developed as part of an explicit and coherent programme of civic renewal, drawing inspiration from the critique of Scottish politics and culture in the Union that came from writers such as Tom Nairn, Michael Fry and Walter Humes.

Three elements could feature in such a programme. Most prominent because potentially most popular would be a radical interpretation of the principles of social justice. That would be able to draw on the firm attachment of people in Scotland to the founding principles of the welfare state, and their equally firm belief in the value of public action. Beyond that, although a minority current, there are radical socialist ideas which have simply not vanished in Scotland. They remain the dominant mood among most intellectuals, and so have a currency through books, pamphlets, novels, plays and poems. They have strong factions of support in both the Labour party and the SNP, probably extending right into the Scottish Government: some half dozen or so of the Labour Ministers had links with the organisation Scottish Labour Action which campaigned for radical home rule and radical social policies in the late 1980s and early 1990s. There are strong social movements with similar beliefs, linked directly to community education. And the evidence of the elections to the Scottish Parliament in May 1999 suggests that there is an electoral base of some 10% which will vote for green or radical socialist parties in a proportional system.

These socialist ideas would not command enough support on their own to sustain a programme of reform. But the sense they create of an intellectual milieu in which radical action is possible will influence debate among more gradual reformers. We look at some examples shortly, but one reason why the radical ideas could permeate the mainstream is the second likely feature of a reforming programme: the sense that the whole point of home rule is to challenge civic Scotland. The problem, it is claimed, is that conservative meliorism has actually pre-empted more radical change. The wider penumbra of socialist ideas tends to ensure that challenging the social democratic consensus does not, in Scotland, lead to a shift to the right, unlike the somewhat analogous challenges made to English civil society by Margaret Thatcher in the 1970s and 1980s. The problem, allegedly, is that comprehensive education did not go far enough, that higher education has expanded with little explicit attention to social justice, and that a respected system of community education has been allowed to atrophy through lack of funds and lack of a systematic legislative base. This inclination to the left does not guarantee popular endorsement of anything resembling socialism, but it does ensure that policy makers are challenged to fulfil the social democratic ideals, not to abandon them.

That is reinforced by the third potential element of a radical programme: an inclination to provoke popular participation against governing groups in civil society. That inclination is common among teachers who are not enamoured of educational managers, school inspectors or national quangos. The greatest potential for radical dissent may in fact be through schisms within civic Scotland, as the apparent civic unity associated with the long campaign for home rule starts to fall apart. Such radicalism was present within that campaigning, as we saw. It become tinged with the philosophy of Gramsci through the writings and teaching of community educators and cultural activists. And so they tend to the view that popular participation will mean radicalisation. Like the consensus-building which can be found in the cautious version of what the Parliament might do, this proposal to construct radical coalitions will not easily mobilise large numbers of people. Its most striking feature, however, will be the constituency of interest groups which it will involve. There would be less of the established lobbyists — the teachers' trade unions, the quangos, the representatives of university principals, for example — and more of the radical dissenters, the kind of activists who dominated discussions on social policy around the fringes of the Labour and nationalist movements in the 1980s. Many of these would be local. That is the basis of the electoral support for the Scottish Socialist Party and the Green Party, each of which won a parliamentary seat in the May 1999 elections. It is the source also of such effective campaigns as the movement for reform of land-ownership in the Highlands, the opposition to school closures, the opposition to standardised testing in primary-school, and — most spectacularly of all — the opposition to the Conservative Government's poll tax from 1989 to 1992. The potency of these campaigns shows that they cannot be ignored as mere fringe activities, and they will have a national platform in the new Civic Forum and the continuing Civic Assembly. Dominated by interest groups these may be, but in Scotland that does not mean uniformly cautious. Some of the interest groups remain anarchic in their radicalism, expressing the frustration of the less politically influential parts of civic Scotland — such as teachers who are not in management — with what they see as a complacent, oppressive and parochial establishment in their own civic backyard.

As with the more cautious future, the implications of this radical one are best seen in how it will deal with each of the four dominant current trends that we assessed in Chapter 4. On expansion, the obvious radical option is to turn the further education colleges into something like the community colleges of the USA, charged explicitly with widening access. Universities would lose money and students to them, unless they too could demonstrate that they were radically broadening their social base. The ultimate aim might even be a form of comprehensive post-school education, by restructuring the first two or three years of higher education into a common format with a common curriculum, structurally the same in universities and colleges. It would draw inspiration not only from the USA but also from a modernised version of the curricular features of the Scottish tradition that was made popular by George Davie — compulsory breadth, and coherence gained through compulsory philosophy. Some of the universities would retain graduate schools, but the initial experience of all students would be common.

That example also illustrates the role which principles of social justice would play. A further instance of that would be the most radical interpretation of the New Community Schools, emphasising their role in providing local cultural leadership. They would draw on the ideals of community schools that were first found in the 1960s, in which the school becomes the focus of community learning, culture and politics. So they would become part of a broader policy of regenerating poor communities, rather than merely a way in which individual pupils could gain the credentials to escape poverty. As one writer put it, commenting on the US full-service schools on which New Community Schools have been modelled:

> considerable attention is being directed toward the involvement of the community and the importance of a sense of ownership by parents and other residents, recalling the language of the community action programs of the 1960s. (Dryfoos, 1995, p. 151)

Indeed, this cultural leadership role for the school would then be relevant to all types of community. In particular, it would link an educational strategy to attempts to regenerate rural areas and to democratise the ownership and control of land. It would thus also radicalise the old Kirriemuir myth which Andrew McPherson described — a reform that would have potent political symbolism.

On standards, a radical programme would not question the allegation that there has been a decline, because that is what would be to be expected from a complacent educational establishment that had subjected itself meekly to 18 years of Conservative rule. Targets would not be set by professionals — or at least not by managing professionals such as head-teachers and inspectors — because they could not be trusted to share the radical vision, and so targets would also go beyond the mere measurement of average examination performance. There would be indicators of equal opportunity — even of average equality of outcome between social groups — and attempts to assess other outcomes such as the capacity of students to think for themselves, and their level of emotional development. How to interpret targets would also not be left to the managers and inspectors, but would be the responsibility of politicians, and of parents on school boards, as well as unpromoted teachers. Democratisation would be taken further than that, drawing inspiration from practice in Denmark and Ireland. Boards would be extended to include representatives of the wider community. Local authorities would have to test their policies in local educational conventions, which would have to include representatives of communities and parents as well as educational interests. Policies for discussion would include not only formal matters such as the organisation and management of schooling, but also matters relating to the curriculum, discipline and child development. A similar structure of broad consultation through a standing convention would be put in place nationally.

Democratisation would also be the inspiration for a programme of citizenship education that encouraged dissent. There would be more emphasis on rights than current ideas in this area tend to favour (as we saw in Chapter 3), and an attention to developing the skills that are required to challenge people in power. This would take place not only in schools but also through community education, where it would be linked to the principle that community educators can and ought to help to bring about social and political change.

These examples of policies are merely sketches, and are certainly not meant to imply that any political group has a radical manifesto available for instant use. They are intended to illustrate the directions in which a radical style of educational politics could emerge in the Parliament. The core idea of all them is that the system needs changing fundamentally, that the existing leadership groups cannot be trusted to do it, and that the only politically acceptable radicalism in Scotland will be of the left. For all the leftist rhetoric, however, another reason why this line of reform could be politically powerful is the convergence of some of these ideas with those coming from other positions. Democratised boards are not that far removed from Conservative party thinking, action on standards would please them and the various right-wing journalists, renewing the democrat intellect would be congenial to nationalists, and scepticism about the teacher unions would even find favour in the Labour party.

Conclusions

The different approaches to education will be tested first of all by how effective they are, and the grand themes will be worked out in the mundane and messy conflicts which dominate the headlines. Three very obvious ones were prominent in the year following the first elections to the Parliament — student finance, teachers' pay, and a bizarre controversy about sex education in schools.

How to pay for the expansion of higher education remained highly controversial in Scotland, and the ostensibly new political process came up with a very old-fashioned way of postponing decisions: a committee of inquiry into one aspect of the problem — student finance — consisting largely of representatives of the main educational interest groups. It did not seem from this instance that politicians were very willing to challenge the civic institutions of education. Evidence to the committee was inevitably dominated by expert witnesses and other educational interest groups (such as various trade unions and the commitee representing higher education principals), and so the outcome was largely what the higher education system itself wanted. That the committee's report (and the diluted version of it which the Scottish government then adopted as policy) was broadly social democratic in tone merely confirms that Scotland's civic institutions are dominated by people on the mildly reformist left. In that sense they are different from analogous bodies in England, but the main point is that it was they who took the lead, not the parliament itself.

On the face of it, the second high-profile conflict did seem to indicate a willingness by the governing coalition to confront civic Scotland: the attempt to impose a radically revised structure of pay and working conditions on school teachers, and the attempt to resolve the resulting impasse by another committee of inquiry, but this time with very little representation from the educational interest that would be most affected (teachers themselves). But taking on teachers is about the easiest stratagem available: they are relatively weak politically, and are reluctant to undertake more than token strikes for fear of harming the education of the children in their care. Indeed, the intention to reform teachers' conditions came, not mainly from outside the education system, but from groups within it who have much more power than teachers: the local authority employers, discretely encouraged by the schools inspectorate. It would have taken far more courage for the Parliament to have challenged them than to challenge the teachers.

Nevertheless, rumbling in the background of this dispute about teachers is the potential for the Parliament to challenge local government's control of school education. That would certainly be a much more radical attack on educational institutions than anything which the Conservative government attempted, although it would also have the effect of vastly strengthening the influence of the most powerful of all, the inspectorate and the national quangos.

The third controversy did reveal parliamentarians willing to challenge parts of civic Scotland, although in the knowledge that most of it was on their side. The controversy had its origins in a piece of legislation that was enacted by the Conservatives in 1988, prohibiting local government from promoting homosexuality as an acceptable family form. Education authority schools, being the responsibility of local government, were caught by this almost inadvertently. When the Scottish Executive proposed to repeal the law, a vociferous campaign of opposition was organised by a millionaire businessman and the Catholic church. These were powerful forces, and defying them required that the Executive emphasise the tolerance which — survey evidence showed — was shared by around three quarters of Scots, rather than the more uncomfortable evidence that a majority were uneasy about repeal. The Parliament supported the Executive's stance by a large majority, and the Conservatives — who were against repeal — refused to indulge in the language that was coming from the other opponents (language that claimed homosexuality to be a 'perversion', for example). Taking on the Catholic church (and ignoring the reservations of the more traditional segments of Scottish left-wing politics, such as the majority on North Lanarkshire Council) did seem to indicate a parliament that was not in awe of at least the more conservative segments of Scotland's civic institutions.

Beyond these most prominent — but probably rather transient — headlines are some deeper issues. Also beginning to creep onto the agenda of debate is a challenge to the national uniformity of provision, at all levels. Higher educational institutions are becoming more diverse, as the funding council encourages specialisation of teaching, of research, and — through targeted rather than blanket schemes to encourage wider access — of social purpose. Nursery education — as we saw in Chapter 4 — is perforce more diverse than primaries and secondaries, because restrictions on public expenditure in the last decade or so have forced local authorities to work in partnership with the private and voluntary sectors. The outcome of this creeping diversification at all levels is impossible to predict, but it is bound to break up the monopoly powers of unions of educational providers. For example, there is already conflict of interest among higher educational institutions, some insisting on funding to maintain their allegedly international status, some giving priority to promoting local access. Disagreements of this sort are one reason why the funding council invites institutions to choose whether to apply for money to promote wider access, but the discretionary character of that then tends to reinforce the diversity of goals.

Not far from the headlines in the past year or so has been the issue of denominational schooling — not just the continued existence of Catholic schools that are fully funded from public sources, while continuing to afford managerial and policy influence to the Catholic church, but also the claims for equal treatment by Muslims and other religious groups. A challenge to the full public funding of separate denominational education would be a threat to one rather important aspect of the civic consensus that has governed Scotland in the Union, and the resulting debates would acrimoniously bring to the fore the question of the right of the Parliament to interfere with civic compromises that seem to have worked to most people's broad satisfaction (or at least with their acquiescence).

It will be in highly visible debates and conflicts of these sorts that the tensions we have been looking at here will surface. Other headlines will appear, but the underlying issues will remain these we have been discussing — expansion, the comprehensive principle, standards and individualism. Running through them all will be questions about the rights and competence of civic Scotland. There is no certainty at all that things will change radically. The cautious reformism may well continue to produce the slow change and slow democratisation which has been achieved since the 1960s (as in the outcome of the debate about student finance). Being cautious, it would not upset anyone and would not risk disruption in the interim. The radical vision might produce striking changes in social inequalities (although the record of this internationally is not good), might draw many more people into making decisions, and might undermine the power of educational vested interests. But both approaches would have drawbacks too: not enough change quickly enough on the one hand, or too much disruption on the other. So there would be plenty of scope for disagreement over how to interpret either route, and so plenty of new headlines.

In any case, neither the education system nor the Parliament is homogeneous, and so both visions will have supporters in both camps. Thus further education colleges are as suspicious of the ancient universities as any radical would be. Teachers are as suspicious of educational bureaucrats (and probably of local authorities and inspectors) as any dissenting politician, whether of the left or the right. And there is certainly not a unanimous civic view about how to teach about sexuality in schools. So civic Scotland is far from being monolithic. The same goes for the Parliament. Despite the apparent unanimity of the first Parliament around cautious reform, the proportional electoral system is bound to erode that in the end. The SNP is likely eventually to evolve a more systematic critique of civic Scotland than it could afford to espouse when the dominant theme of Scottish politics was defending Scottish achievements. A rejuvenated Conservative party that will have become firmly Scottish in its cultural orientation will be in a much better position to question social democratic caution than Micheal Forsyth could achieve from his position heading an embattled minority government. If the Labour party does not reform civic Scotland — especially its own backyard in the local authorities — then it will entice the SNP as well as the Conservatives into a more radical critique of civic complacency.

And lurking around the fringes will be the smaller parties — the Liberal Democrats with a tinge of rural radicalism, and the Scottish Socialists and Greens who have no reason to be enamoured of either the Scottish Government or the traditional governing groups in civic Scotland.

In this new context, the defensive unity of the education system that was provoked by Thatcher will gradually decay, and the radical currents within it will find themselves as much in opposition to educational leaders as to the Government or the local authorities. That happened in the 1960s: comprehensive education went ahead smoothly because there was a large minority of teachers who supported it, despite the reservations of their own trade unions and many prominent educationalists. The very fact that education is relatively autonomous would allow radical teachers to take their own action. Community educators are independent enough to allow them to be explicit about seeing their role as helping to bring about social change. Staff in universities have the freedom to write and speak in favour of radical action, despite the views of their principals. Even school teachers, tightly constrained though they now are, have a great deal of freedom to subvert national policy, as we saw in Chapter 3. For example, in their own classrooms, they do not have to be as obsessed with examination performance as the inspectorate's target-setting exercise seems to require.

Radical educators of this sort would be more adventurous than the cautious majority of members of the Parliament, and so would claim that they — not the MSPs — were the true inheritors of the idealism which led to the setting up of the Parliament. That would be a version of the claims that the education system knew better what to do than the politicians. It would not be a conservative defence but a radical challenge, but it would be able to borrow some of the rhetoric about the system's autonomy from political interference. And then radical MSPs might champion their cause, especially if the Scottish Government seems mired in caution. In particular, there might then be some political inclination to question the role of the inspectorate and its network of quangos, something which was almost absent from public political debate in the unreformed Union.

Nevertheless, there is nothing inevitable about that, or even self-evidently anything very desirable. The power of the cautious educational institutions could prevail, the defensiveness that is endemic in any large profession could dominate thinking among teachers, and the radical critics could appear to be a mixture of leftist cranks, discredited Tories, and nationalists with ulterior motives. And there may appear to be enough validity in all these points to persuade the majority that the best way forward is cautious reform, especially since there are indeed powerful tendencies in the new Parliament to avoid confrontation. The reformers would borrow some of the rhetoric of the radicals, to show that they had not betrayed the idealism, and they would adapt some of the radicals' ideas, just as the instigators of comprehensive education did in the 1960s. For example, they would use citizenship education to encourage some types of participation, in the styles of consultation that were sanctioned by the Parliament. They would justify the expansion of further education colleges partly in terms of widening access. And they would take advantage of the spectrum of opinion on New Community Schools, going further than merely re-arranging local services under one roof, but stopping well short of encouraging schools to lead the community in any political challenge to central authorities. As throughout the twentieth century, the reformers would also combine the radicals' idealism with the human capital justification for expansion. More higher education, it would be claimed, would not only be good for the economy but would also liberate individuals by realising their full potential.

All of this seems far removed from the apparently widespread belief that education would automatically and uncontroversially benefit from a Scottish Parliament. The argument presented in this book has suggested that deep divisions remain, because none of the visions that are offered is a straightforward means to educational improvement. Neither working with the cautious reformism of civic institutions nor challenging their very legitimacy will be bound to achieve fundamental reform, and there is not even agreement on how much reform is needed, nor to what ultimate purpose. But, also, neither of these strategies would be disastrous. The Scottish civic institutions are not as conservative as their critics allege, because — as we have seen — they have presided over gradual reform when forced to do so by social change and discrete political pressure. And radical action, although disruptive, would be moderated in practice by the unavoidable caution of teachers and civic leaders, just as it was in the 1960s and earlier.

That ambiguous conclusion means simply that Scottish politics is becoming normal. The tension between civic institutions and national parliament is now contained within Scotland's own polity, just as in most other countries: no longer is it a conflict between an indigenous civil society and an alien state. So the resolution of the tensions in Scotland will involve the same kinds of compromises and victories as in other places. Utopian simplicity is no longer tenable, as it was when the Parliament was still a dream. The rule of the civic elites is no longer impregnable, as it was when straightforward Unionism dominated Scottish culture. The existence of the Parliament forces a more open debate between these positions, and so helps to make the discussion of policy more realistic but also more open to innovative ideas. There is no easy resolution of all the many currents of thinking which have been traced here, but their very strength and complexity ensure that the politics of Scottish education is starting to become creative.

FURTHER READING

Chapter 1

The general debate on Scottish self-government can be followed in Brown et al (1998), Devine (1999), Mitchell (1996), Paterson (1994, 1998b) and Taylor (1999). The connection between education and nationalism is discussed by Gellner (1983), Green (1997), the essays in Schleicher (1993) and — for Ireland — Lyons (1973) and Brown (1985); for Scotland, the nationalism lying in the background of most educational policy making is assessed by Paterson (1997a). The role of the state is covered by many authors, including Archer (1979) and Green (1997). The changes which a mass system brought about are dealt with by them, and by McPherson (1993), Ramirez and Boli (1987), Boli, Ramirez and Meyer (1985) and Simon (1974, 1991). The role of civil society in renewing the state is discussed by, for example, Cohen and Arato (1992), Hearn (2000), Keane (1998) and Putnam (1993).

Chapter 2

General histories of Scottish education used to be plentiful; since the 1960s, there has been a dearth, especially relating to the twentieth century, despite the very valuable historical work done by R. D. Anderson and D. J. Withrington. Some general texts are Anderson (1983, 1995), Gray et al (1983), Houston (1985), McPherson (1993), McPherson and Raab (1988), Osborne (1968), Paterson (1996, 1998c), Scotland (1969) and Withrington (1983, 1988). The essays in Clark and Munn (1998) and Bryce and Humes (1999a) provide authoritative accounts of current policy and practice, with some historical material. The books by Davie (1961, 1986), although hotly contested, have shaped the entire debate recently about the significance of the Scottish educational tradition. All these works provide the main sources for this chapter. On more specific topics that are not well covered in the general texts, see Crowther et al (1999) for community education, Alexander et al (1995) for further education colleges, and Curtice and Jowell (1997) and Jowell and Topf (1988) for trust in professionals in many countries.

Chapter 3

The general history of the movement for Scottish home rule has been told in many places, as noted above for Chapter 1. The role of quangos in governing complex public services such as education was the subject of a special issue of the journal *Parliamentary Affairs* (1995, volume 48, number 2). The attempts to develop a more open style of policy making in Scotland are found in Crick and Millar (1995), the Advisory Committee on Telematics (1997) and the report of the Consultative Steering Group (Scottish Office, 1998). Citizenship education is mentioned in these reports, and is discussed more fully by Carr (1991), Avis et al (1996), Hall and Held (1989), Pierson (1991) and the report of a committee chaired by Crick for the English Qualifications and Curriculum Authority (1998). Some of the distinctively Scottish aspects of citizenship are dealt with by Morton (1998) and Paterson (2000b). The potential role of women and the women's movement in changing the character of Scottish democracy is discussed in Chapter 8 of Brown et al (1998). Thinking skills are discussed by Lipman (1993) and Nisbet (1993); the relevance of emotional intelligence and other such topics to educational reform is argued by Scottish Council Foundation (1999). The main sources for a critique of civic Scotland are Nairn (1997), Humes (1986) and Fry (1987).

Chapter 4

The recent expansion of Scottish education is traced by Paterson (1997b), and some aspects of the distinctiveness of the system are discussed by Paterson (2000). The expansion of the proportion passing Higher English was calculated from the Scottish School Leavers Survey (described by, for example, Gray et al (1983) and Lynn (1994)). For the more general aspects of the so-called knowledge revolution, of growing individualism, and of growing popular desires to participate in decision making, see Bentley (1998), Giddens (1994, 1999), Inglehart (1990) and Ryan (1999). Comparisons between Scotland and England in this regard are made by Miller et al (1996, pp. 369–73). The failure of the social democratic left to respond to growing individualism is dealt with by Giddens (1994), Sassoon (1996), Taylor-Goodby (1989), Wainwright (1994) and Hall (1989), and the new left response can be found in Hall (1989), London Edinburgh Weekend Return Group (1979), Rowbotham et al (1979) and Sassoon (1996). The prominent role of new social movements is dealt with by Eder (1993), Della Porta and Diani (1999) and — for Scotland — Crowther et al (1999). The Youth Congress has been described in the official publication *What's Happening in the Scottish Parliament* (for example, 6 November 1999, p. 27).

The fate of the comprehensive principle can be found in Paterson (1998d) and Benn and Chitty (1996). The cultural role of comprehensive education as developed in Scotland is assessed by Kirk (1986) and by the essays in Kirk and Glaister (1994). Doubts about the economic effects of expansion are summarised by Ashton and Green (1996). The sieving function of education is dealt with by several of the chapters in Halsey et al (1997), notably by the Goldthorpe (1996) paper which is reprinted there. The effect of expansion in stimulating further demand is examined by Burnhill et al (1990). School ethos is the subject of Munn (1999a, b). The right-wing critique of comprehensive and child-centred education is found in, for example, Peters (1969); the more recent appearance of this in Scottish debates is commonly found in the pages of *The Scotsman* in the last few years, for example Luckhurst (1997). The leftist and nationalist version of the critique of standards is exemplified in the pamphlet by the Scottish Association of Teachers of Language and Literature (1999). The distinctiveness of Scottish approaches to child-centredness is assessed by Darling (1994, 1999). The results of the Assessment of Achievement Programme are exemplified in Macnab et al (1989), and the long-running controversies can be found in the pages of the *Times Education Supplement Scotland*, for example Munro (1988, 1999). The professional culture of Scottish teachers is covered by Kirk (1988).

The operation of devolved school management was investigated by Adler et al (1997). New Community Schools were proposed by the SOEID (1998), modelled on full-service schools in the USA (Dryfoos, 1995). Early intervention was also pioneered by the Scottish Office (SEED, 1999). Reforms of community education are proposed by (SOEID, 1999a). The guru of communitarianism is Etzioni (1993).

Chapter 5

The aspiration to create a new style of politics in and around the Scottish Parliament is discussed in Chapter 5 of Brown et al (1998). The Irish Education Convention is reported by Coolahan (1994). Aspects of policy making in Danish education are dealt with by DES (1988), by various reports from the Danish Ministry of Education (1996, 1997a,b) and by some of the essays in Allardt (1981). The power of the Scottish inspectorate is criticised by Bryce and Humes (1999b); Michael Forsyth's failure to do much about that is dealt with by Humes (1995). Attitudes to the Labour minister Helen Liddell are reported in the *Herald* (8 August 1998, p. 60), for the SNP, and the *Herald* (18 January 1999, p. 1), for the EIS. More general attacks on allegedly anglicising tendencies in the Scottish Labour government can be found in several pamphlets, speeches and articles by teachers: for example, Scottish Association of Teachers of Language and Literature (1999). The radical and Gramscian influences on community education are evident in Crowther et al (1999). The report of the committee of inquiry into student finance is available on the web at http://www.studentfinance.org.uk.

REFERENCES

Adler, M., Arnott, M., Bailey, L., McAvoy, L., Munn, P. and Raab, C. (1997), *Devolved Management in Secondary Schools in Scotland*, Edinburgh: Department of Politics, Edinburgh University.

Advisory Committee on Telematics for the Scottish Parliament (1997), *A Parliament for the Millenium*, Edinburgh: John Wheatley Centre.

Alexander, H., Gallacher, J., Leahy, J. and Yule, B. (1995), 'Changing patterns of higher education in Scotland: a study of links between further education colleges and higher education institutions', *Scottish Journal of Adult and Continuing Education*, 2, 25–44.

Anderson, B. (1983), *Imagined Communities: Reflections on the Origins and Spread of Nationalism*, London: Verso.

Allardt, E (ed.) (1981), *Nordic Democracy: Ideas, Issues, and Institutions in Politics, Economy, Education, Social and Cultural affairs of Denmark, Finland, Iceland, Norway, and Sweden*, Copenhagen: Det Danske Selskab.

Anderson, R. D. (1983). *Education and Opportunity in Victorian Scotland*, Edinburgh: Edinburgh University Press.

Anderson, R. D. (1992), 'The Scottish university tradition: past and future', in J. Carter and D. Withrington (eds), *Scottish Universities: Distinctiveness and Diversity*, Edinburgh, 67–78.

Anderson, R. D. (1995). *Education and the Scottish People, 1750–1918*. Oxford: Oxford University Press.

Archer, M. (1979), *Social Origins of Educational Systems*, London: Sage.

Ashton, D. and Green, F. (1996), *Education, Training and the Global Economy*, Cheltenham: Edward Elgar.

Avis, J., Bloomer, M., Esland, G., Gleeson, D. and Hodkinson, P. (1996), *Knowledge and Nationhood*, London: Cassell.

Benn, C. and Chitty, C. (1996), *Thirty Years On*, London: David Fulton.

Bentley, T. (1998), *Learning Beyond the Classroom*, London: Routledge.

Bloomer, M. (1996), 'Education for citizenship', in Avis et al (1996), 140–63.

Boli, J., Ramirez, F. O. and Meyer, J. W. (1985), 'Explaining the origins and expansion of mass education', *Comparative Education Review*, 29, 145–70.

Boyd, B. (1994), 'The management of curriculum development: the 5–14 programme', in W. M. Humes and M. L. Mackenzie (eds), *The Management of Educational Policy*, Harlow: Longman, 17–30.

Brown, A., McCrone, D., Paterson, L. (1998), *Politics and Society in Scotland*, London: Macmillan, second edition.

Brown, A., McCrone, D., Paterson, L. and Surridge, P. (1999), *The Scottish Electorate*, London: Macmillan.

Brown, S. and McIntyre, D. (1993), *Making Sense of Teaching*, Buckingham: Open University Press.

Brown, T. (1985), *Ireland: a Social and Cultural History, 1922–1985*, London: Fontana.

Bryce, T. G. K. and Humes, W. M. (1999a), *Scottish Education*, Edinburgh: Edinburgh University Press.

Bryce, T. G. K. and Humes, W. M. (1999b), *Policy Development in Scottish Education*, Glasgow: Universities of Glasgow and Strathclyde.

Burnhill, P., Garner, C. and McPherson, A. (1990), 'Parental education, social class and entry to higher education, 1976–1986', *Journal of the Royal Statistical Society*, series A, 153, 233–48.

Carr, W. (1991), 'Education for citizenship', *British Journal of Educational Studies*, 39, 373–85.

Cipolla, C. M. (1969), *Literacy and Development in the West*, Harmondsworth: Penguin.

Clark, M. and Munn, P. (eds) (1998), *Education in Scotland*, London: Routledge.

Cohen, J. L. and Arato, A. (1992), *Civil Society and Political Theory*, Cambridge, Mass.: MIT Press.

Committee on Higher Education (1963), *Higher Education: Appendix One: the Demand for Places in Higher Education*, London: HMSO.

Coolahan, J. (1994), *Report on the National Education Convention*, Dublin: National Education Convention.

Crick, B. and Millar, D. (1995), *To Make the Parliament of Scotland a Model for Democracy*, Edinburgh: John Wheatley Centre.

Crowther, J., Martin, I. and Shaw, M. (1999). *Popular Education and Social Movements in Scotland Today*. Leicester: National Institute of Adult and Continuing Education.

Curtice, J. and Jowell, R. (1997), 'Trust in the political system', in R. Jowell, J. Curtice, A. Park, L. Brook, K. Thomson and C. Bryson (eds), *British Social Attitudes: the 14th Report*, Aldershot: Ashgate, 89–109.

Daiches, D. (1977), *Scotland and the Union*, London: J. Murray.

Danish Ministry of Education (1996), *Act on the Folkeskole*, Copenhagen: Ministry of Education.

Danish Ministry of Education (1997a), *Principles and Issues in Education*, Copenhagen: Ministry of Education.

Danish Ministry of Education (1997b), *Administration of the Folkeskole*, Copenhagen: Ministry of Education.

Darling, J. (1994), *Child-Centred Education and its Critics*, London: Paul Chapman.

Darling, J. (1999), 'Scottish primary education: philosophy and practice', in Bryce and Humes (eds), 27–36.

Davie, G. E. (1961), *The Democratic Intellect: Scotland and Her Universities in the Nineteenth Century*, Edinburgh: Edinburgh University Press.

Davie, G. E. (1986), *The Crisis of the Democratic Intellect: the Problem of Generalism and Specialism in Twentieth-Century Scotland*, Edinburgh: Edinburgh University Press.

Della Porta, D. and Diani, M. (1999), *Social Movements*, Oxford: Blackwell.

Department of Education and Science (1988), *Education in Denmark*, London: DES.

Derry, T.K. (1973), *A History of Modern Norway, 1814–1972*, Oxford: Clarendon.

Devine, T. M. (1999), *The Scottish Nation*, Harmondsworth: Penguin.

Dryfoos, J.G. (1995), 'Full service schools: revolution or fad?', *Journal of Research on Adolescence*, 5, 147–72.

Dyson, A. E. and Lovelock, J. (1975), *Education and Democracy*, London: Routledge and Kegan Paul.

Eder, K. (1993), *The New Politics of Class*, London: Sage.

Educational Institute of Scotland (1999), *Manifesto for a New Parliament*, Edinburgh: EIS.

Edwards, O. D. (ed.) (1989), *A Claim of Right for Scotland*, Edinburgh: Polygon.

Erikson, R. and Jonsson, J. O. (eds) (1996), *Can Education be Equalised?*, Oxford: Westview.

Etzioni, A. (1993), *The Spirit of Community*, New York: Simon and Schuster.

Flett, I. (1989), *Association of Directors of Education in Scotland: the Years of Growth, 1945–1975*, ADES.

Fry, M. (1987), *Patronage and Principle: a Political History of Modern Scotland*, Aberdeen: Aberdeen University Press.

Gellner, E. (1983), *Nations and Nationalism*, Oxford: Blackwell.

Giddens, A. (1994), *Beyond Left and Right*, Cambridge: Polity.

Giddens, A. (1999), *The Third Way*, Cambridge: Polity.

Goldthorpe, J. (1996), 'Problems of "meritocracy"', in Erikson and Jonsson (1996) (eds), 255–87; reprinted in Halsey et al (1997), 663–82.

Gray, J., McPherson, A. and Raffe, D. (1983), *Reconstructions of Secondary Education: Theory, Myth and Practice Since the War*, London: Routledge and Kegan Paul.

Green, A. (1990), *Education and State Formation: the Rise of Education Systems in England, France and the USA*, London: Macmillan.

Green, A. (1997), *Education, Globalisation and the Nation State*, London: Macmillan.

Grieve, C. M. (1926), 'A. S. Neill and our educational system', *Scottish Educational Journal*, 5 March, 241–3.

Hall, S. (1989), 'The meaning of New Times', in Hall and Jacques (eds), 116–36.

Hall, S. and Held, D. (1989), 'Citizens and citizenship', in Hall and Jacques (eds), 173–88.

Hall, S. and Jacques, M. (eds) (1989), *New Times*, London: Lawrence and Wishart.

Halpern, D. (1995), 'Values, morality and modernity: the values, constraints and norms of European youth', in M. Rutter and D. J. Smith (eds), *Pscychosocial Disorders in Young People*, New York: Wiley, 324–87.

Halsey, A.H., Lauder, H., Brown, P. and Wells, A.S. (eds.) (1997), *Education: Culture, Economy, Society*, Oxford: Oxford University Press.

Hearn, J. (2000), "Introduction", in J. Hearn (ed.), *Taking Liberties: Contesting Visions of the Civil Society Project*, special issue of *Critique of Anthropology*.

Hills, L. (1990), 'The Senga syndrome: reflections on 21 years in education', in F. M. S. Paterson and J. Fewell (eds), *Girls in Their Prime*, Edinburgh: Scottish Academic Press, 148–66.

Houston, R. A. (1985), *Scottish Literacy and Scottish Identity, 1600–1800*, Cambridge: Cambridge University Press.

Humes, W. M. (1986), *The Leadership Class in Scottish Education*, Edinburgh: John Donald.

Humes, W. (1995), 'The significance of Michael Forsyth in Scottish education', *Scottish Affairs*, no. 11, 112–30.

Humes, W. M. and Paterson, H. M. (1983) (eds), *Scottish Culture and Scottish Education*, Edinburgh: John Donald.

Hutton, W. (1995), *The State We're In*, London: Jonathan Cape.

Inglehart, R. (1990), *Culture Shift in Advanced Industrial Society*, Princeton University Press.

Jowell, R. and Topf, R. (1988), 'Trust in the establishment', in R. Jowell, S. Witherspoon and L. Brook (eds), *British Social Attitudes: the 5th Report*, Aldershot: Gower, 109–26.

Keane, J. (1998), *Civil Society*, Cambridge: Polity.

Kirk, G. (1986), *The Core Curriculum*, London: Hodder and Stoughton.

Kirk, G. (1988), *Teacher Education and Professional Development*, Edinburgh: Scottish Academic Press.

Kirk, G. and Glaister, R. (1994), *5–14: Scotland's National Curriculum*, Edinburgh: Scottish Academic Press.

Lee, J. (1963), *This Great Journey: a Volume of Autobiography, 1904–1945*, London: MacGibbon and Kee.

Lindsay, I. (1976), 'Nationalism, community and democracy', in G. Kennedy (ed.), *The Radical Approach*, Edinburgh: Palingenesis Press, 21–6.

Lipman, M. (1993), 'Promoting better classroom thinking', *Educational Psychology*, 13, 291–304.

London Edinburgh Weekend Return Group (1979), *In and Against the State*, London: Pluto Press.

Luckhurst, T. (1997), 'Selection encourages academic excellence', *The Scotsman*, 14 February, 18.

Lynn, P. (1994), *The 1994 Leavers*, Edinburgh: SOEID.

Lyons, F. S. L. (1973), *Ireland since the Famine*, London: Fontana.

Mackintosh, J. P. (1974), 'The new appeal of nationalism', *New Statesman*, 27 September, 408–12; reprinted in Paterson (ed.) (1998a), 64–71.

Mackintosh, J. P. (1975), *A Parliament for Scotland*, Tranent: Berwick and East Lothian Labour Party; reprinted in Paterson (ed.) (1998b), 83–7.

Macnab, D., Page, J. and Kennedy, M. (1989), *Assessment of Achievement Programme: Second Round 1988, Mathematics*, Aberdeen: Northern College.

Marshall, T. H. (1950), *Citizenship and Social Class and Other Essays*, Cambridge: Cambridge University Press.

Maxton, J. (1926), 'A national aim in education', *Scottish Educational Journal*, 8 January, 37–9.

McPherson, A. (1983). 'An angle on the geist: persistence and change in the Scottish educational tradition', in W. M. Humes and H. M. Paterson (eds) *Scottish Culture and Scottish Education*, Edinburgh: John Donald, 216–43.

McPherson, A. (1993). 'Schooling', in A. Dickson and J. H. Treble (eds), *People and Society in Scotland, vol III, 1914–1990*, Edinburgh: John Donald, 80–107.

McPherson, A. and Raab, C. D. (1988), *Governing Education*, Edinburgh: Edinburgh University Press.

McPherson, A. and Willms, J. D. (1987), 'Equalisation and improvement: some effects of comprehensive reorganisation in Scotland', *Sociology*, 21, 509–39.

Miller, W. L., Timpson, A. M. and Lessnoff, M. (1996), *Political Culture in Contemporary Britain*, Oxford: Clarendon.

Mitchell, J. (1996), *Strategies for Self-Government*, Edinburgh: Polygon.

Morton, G. (1998), 'Civil society, municipal government and the state: enshrinement, empowerment and legitimacy. Scotland, 1800–1929'. *Urban History*, 25, 348–67.

Morton, G. (1999). *Unionist Nationalism*, East Linton: Tuckwell.

Munn, P. (1999a), 'Ethos and discipline in the secondary school', in Bryce and Humes (eds), 406–14.

Munn, P. (1999b), 'Research and practice', in Bryce and Humes (eds), 952–60.

Munro, N. (1988), 'Forsyth says "disturbing" results justify testing', *Times Educational Supplement Scotland*, 16 December, 3.

Munro, N. (1999), 'AAP alert on literacy', *Times Educational Supplement Scotland*, 12 November, 1.

Nairn, T. (1970), 'The three dreams of Scottish nationalism', in K. Miller (ed.), *Memoirs of a Modern Scotland*, London: Faber, 34–54; part reprinted in Paterson (ed.) (1998b), 31–39.

Nairn, T. (1997), *Faces of Nationalism*, London: Verso.

Nisbet, J. (1993), 'The thinking curriculum', *Educational Psychology*, 13, 281–89.

Osborne, G. S. (1968), *Change in Scottish Education*, London: Longmans Green.

Paterson, H. M. (1983), 'Incubus and ideology: the development of secondary schooling in Scotland, 1900–1939', in Humes and Paterson (eds), 14–21.

Paterson, L. (1994). *The Autonomy of Modern Scotland*, Edinburgh: Edinburgh University Press.

Paterson, L. (1996), 'Liberation or control: what are the Scottish educational traditions in the twentieth century?', in T. M. Devine and R. J. Finlay (eds), *Scotland in the Twentieth Century*, Edinburgh University Press, 230–49.

Paterson, L. (1997a), 'Policy making in Scottish education: a case of pragmatic nationalism', in P. Munn and M. Clark (eds), *Education in Scotland*, London: Routledge, 138–55.

Paterson, L. (1997b), 'Student achievement and educational change in Scotland, 1980–1995', *Scottish Educational Review*, 29, 10–19.

Paterson, L. (1998a), 'The civic activism of Scottish teachers: explanations and consequences', *Oxford Review of Education*, 24, 279–302.

Paterson, L. (ed.) (1998b), *A Diverse Assembly: the Debate on a Scottish Parliament*, Edinburgh: Edinburgh University Press.

Paterson, L. (1998c), 'Scottish higher education and the Scottish parliament: the consequences of mistaken national identity', *European Review*, 6, 459–74.

Paterson, L. (1998d), 'Education, local government and the Scottish parliament', *Scottish Educational Review*, 29, 52–60.

Paterson, L. (2000a), 'Does civil society speak for the people? Evidence from a survey of Scottish teachers', *Sociological Review*, to appear.

Paterson, L. (2000b), 'Civil society and democratic renewal', in S. Baron, J. Field, and T. Schuller (eds), *Social Capital: Social Theory and the Third Way*, Oxford: Oxford University Press, to appear.

Paterson, L. (2000c), 'Scottish traditions in education', in H. Holmes (ed.), *Compendium of Scottish Ethnology, vol 11*, Edinburgh: Scottish Ethnological Research Centre, to appear.

Paton, H. J. (1968), *The Claim of Scotland*, London: George Allen and Unwin.

Peters, R. S. (1969), *Perspectives on Plowden*, London: Routledge and Kegan Paul.

Pierson, C. (1991), *Beyond the Welfare State?*, Cambridge: Polity.

Preston, P. (1999), 'Keep your eyes wide open in the schooling debate', *Guardian*, 23 August, 24.

Putnam, R.D. (1993), *Making Democracy Work.*, Princeton: Princeton University Press.

Qualifications and Curriculum Authorty (1998), *Education for Citizenship and the Teaching of Democracy in Schools*, Final Report of the Advisory Group on Citizenship (chaired by Bernard Crick), London: QCA.

Ramirez, F. O. and Boli, J. (1985), 'The political construction of mass schooling: European origins and worldwide institutionalisation', *Sociology of Education*, 60, 2–17.

Ranson, S. (1994), *Towards the Learning Society*, London: Cassell.

Rowbotham, S., Segal, L. and Wainwright, H. (1979), *Beyond the Fragments*, London: Merlin.

Ryan, A. (1999), *Liberal Anxieties and Liberal Education*, London: Profile.

Sassoon, D. (1996), *One Hundred Years of Socialism*, London: I. B. Tauris.

Schleicher, K. (ed.) (1993), *Nationalism in Education*, Frankfurt: Peter Lang.

Scotland, J. (1969). *The History of Scottish Education (two volumes)* London: University of London Press.

Scottish Association of Teachers of Language and Literature (1999), *Sense and Worth*, Buckstone: SATOLL.

Scottish Centre for Economic and Social Research (1989), *Scottish Education: a Declaration of Principles*, Edinburgh: Scottish Centre for Economic and Social Research.

Scottish Conservative and Unionist Party (1998), *Scotland's Future*, Edinburgh: Scottish Conservative and Unionist Party.

Scottish Consultative Council on the Curriculum (1999), *The School Curriculum and the Culture of Scotland*, Dundee: Scottish Consultative Council on the Curriculum.

Scottish Council Foundation (1999), *Children, Families and Learning*, Edinburgh: Scottish Council Foundation.

Scottish Education Department (1947), *Secondary Education*, Edinburgh: SED.

Scottish Education Department (1975), *Adult Education: the Challenge of Change*, Edinburgh: SED.

Scottish Education Department (1977), *Nursery Education*, Statistical Bulletin, no.4/A2/1977, Edinburgh: HMSO.

Scottish Education Department (1991), *School Leavers' Qualifications 1988–89*, Statistical Bulletin, Edn/E2/1991/4, Edinburgh: HMSO.

Scottish Executive (1999a), *Statistics on Students in Higher Education in Scotland: 1997–98*, Edinburgh: Scottish Executive.

Scottish Executive (1999b), *Improving our Schools*, Edinburgh: Scottish Executive.

Scottish Executive Education Department (1999), *Early Intervention*, Edinburgh: SEED.

Scottish Labour Party (1999), *Building Scotland's Future*, Glasgow: Scottish Labour Party.

Scottish National Party (1999), *Manifesto for the Scottish Parliament 1999 Elections*, Edinburgh: SNP.

Scottish Office (1998), *Shaping Scotland's Parliament*, Edinburgh: Scottish Office.

Scottish Office (1999), *Moving Forward: Local Government and the Scottish Parliament*, Edinburgh: Scottish Office.

Scottish Office Education and Industry Department (1996), *Standards and Quality in Scottish Schools, 1992–95*, Edinburgh: Audit Unit.

Scottish Office Education and Industry Department (1995), *Provision for Pre-School Education*, Statistical Bulletin, Edn/A2/1995/16, Edinburgh: HMSO.

Scottish Office Education and Industry Department (1997), *Achievements of Primary 4 and Primary 5 Pupils in Mathematics and Science*, Edinburgh: SOEID.

Scottish Office Education and Industry Department (1998a), *School Leavers and their Qualifications, 1986–7 to 1996–7*, Statistical Bulletin, Edn/E2/1998/6, Edinburgh: HMSO.

Scottish Office Education and Industry Department (1998b), *New Community Schools Prospectus*, Edinburgh: SOEID.

Scottish Office Education and Industry Department (1999a), *Communities: Change Through Learning*, Edinburgh: SOEID.

Scottish Office Education and Industry Department (1999b), *Standards and Quality in Scottish Schools, 1995–1998*, Edinburgh: Audit Unit.

Scottish Office Education Department (1994), *School Leavers and their Qualifications, 1982–3 to 1992–3*, Statistical Bulletin, Edn/E2/1994/9, Edinburgh: HMSO.

Simon, B. (1974), *The Politics of Educational Reform, 1920–1940*, London: Lawrence and Wishart.

Simon, B. (1991), *Education and the Social Order, 1940–1990*, London: Lawrence and Wishart.

Smith, J. (1981), 'Portrait of a devolutionist: interview with John Smith', *Bulletin of Scottish Politics*, 2, 44–54; reprinted in Paterson (ed.) (1998b), 135–40.

Smout, T. C. (1986), *A Century of the Scottish People, 1830–1950*, London: Collins.

Stocks, J. (1995), 'The people versus the department: the case of Circular 44', *Scottish Educational Review*, 27, 48–60.

Stow, D. (1847), *National Education: the Duty of England in Regard to the Moral and Intellectual Elevation of the Poor and Woking Classes*, London: J.Hatchard and Son.

Swanson, A.M.M. (1975), *The History of Edinburgh's Nursery Schools*, Edinburgh: British Association for Early Childhood Education.

Taylor, B. (1999), *The Scottish Parliament*, Edinburgh: Polygon.

Taylor-Goodby, P. (1989), 'The role of the state', in R. Jowell, S. Witherspoon, and L. Brook (eds), *British Social Attitudes: Special International Report*, Aldershot: Gower, 35–58.

Turner, R. H. (1960), 'Sponsored and contested mobility and the school system', *American Sociological Review*, 25, 855–67.

Wainwright, H. (1994), *Arguments for a New Left*, Oxford: Blackwell.

Watt, J. (1990), *Early Education: the Current Debate*, Edinburgh: John Donald.

Williams, R. (1961), *The Long Revolution*, Harmondsworth: Penguin.

Withrington, D. J. (1983), '"Scotland a half educated nation' in 1834? Reliable critique or persuasive polemic?", in W. M. Humes and H. M. Paterson (eds), *Scottish Culture and Scottish Education, 1800–1980*, Edinburgh: John Donald, 55–74.

Withrington, D. J. (1988a), 'Schooling, literacy and society', in T. M. Devine and R. Mitchison (eds), *People and Society in Scotland, vol I, 1760–1830*, Edinburgh: John Donald, 163–87.

Withrington, D. J. (1988b). '"A ferment of change": aspirations, ideas and ideals in nineteenth-century Scotland', in D. Gifford (ed.), *The History of Scottish Literature: Volume 3, Nineteenth Century*. Aberdeen: Aberdeen University Press, 43–63.